BIOHACKER'S GUIDE TO PEPTIDE PROTOCOLS

Reveal the Secrets to Ageless Vitality, Faster Muscle Recovery, Deep Sleep, and Sharper Focus

Linden Hayes

Part 1:
Fundamentals of Peptide Science

© 2024 Linden Hayes. All rights reserved.

This document is intended exclusively for informational purposes in connection with 'Biohacker's Guide to Peptide Protocols.' Unauthorized copying, sharing, or dissemination of this book, in full or in part, is strictly forbidden. The publisher assumes no responsibility for any harm or loss arising from the application or misapplication of the information contained within these pages. This book is provided 'as is' without any guarantees, whether expressed or implied, regarding its content or utility. All trademarks and brand names cited herein are the exclusive property of their respective owners. Readers are encouraged to conduct their own research and consult with healthcare professionals before implementing any protocols or recommendations found in this book. The information presented is not intended to replace professional medical advice, diagnosis, or treatment. The publisher and authors do not endorse any specific products, treatments, or companies mentioned. Use of the information in this book is at the reader's own risk. By accessing this book, you acknowledge and agree to these terms and accept full responsibility for any consequences resulting from its use. Please respect the intellectual property rights and refrain from unauthorized use

Part 1:
Fundamentals of Peptide Science

TABLE OF CONTENT

PART 1: FUNDAMENTALS OF PEPTIDE SCIENCE 7

CHAPTER 1: PEPTIDES 101 8
    1.1: What Are Peptides? 10
    1.2: Peptides at the Cellular Level 12
    1.3: Natural vs. Synthetic Peptides 13
    1.4: Peptides for Health Optimization 15

CHAPTER 2: QUALITY, SAFETY, AND SOURCING ESSENTIALS 18
    2.1: Ensuring Peptide Quality 18
    2.2: Peptide Purity and Potency 20
    2.3: Peptides Legality and Regulations 21
        2.3.1: Understanding US Peptide Regulations 22
        2.3.2: Global Legal Landscape for Peptide Use 24
        2.3.3: Safe and Legal Peptide Purchase Tips 25

CHAPTER 3: PRACTICAL GUIDELINES FOR PEPTIDES 28
    3.1: Determining the Right Dosage 28
        3.1.1: Calculating Dosages by Weight and Goals 30
        3.1.2: Dosage Guidelines for Beginners vs. Advanced 31
    3.2: Safe Injection Practices and Storage Tips 33
        3.2.1: Peptide Storage and Handling Guidelines 34
        3.2.2: Subcutaneous vs. Intramuscular Injections 35
        3.2.3: Avoiding Injection Contamination Mistakes 37
    3.3: Cycling and Timing Peptides for Results 39

PART 2: TARGETED HEALTH BENEFITS WITH PEPTIDES 41

CHAPTER 4: PEPTIDES FOR FAT LOSS AND METABOLISM 42
    4.1: Peptides for Fat Burning and Metabolism 45
    4.2: Key Peptides for Fat Loss 47
        4.2.1: AOD-9604 Fat Reduction Protocols 48

  4.2.2: CJC-1295 and Ipamorelin Synergy 50

  4.3: Protocols for Effective Weight Management 51

CHAPTER 5: MUSCLE BUILDING WITH PEPTIDES 54

  5.1: Muscle Synthesis and Recovery Mechanisms.. 57

  5.2: Top Peptides for Muscle Growth... 58

    5.2.1: GHRP-2 and GHRP-6 for Muscle Growth 59

    5.2.2: IGF-1 LR3 for Muscle Enhancement 61

  5.3: Peptide Protocols for Accelerated Recovery 62

    5.3.1: Peptide Stacking for Muscle Repair 64

    5.3.2: Enhancing Recovery for High-Intensity Training................................... 65

CHAPTER 6: PEPTIDES FOR COGNITIVE CLARITY 68

  6.1: Enhancing Focus and Memory with Peptides 71

  6.2: Essential Peptides for Cognitive Function 72

    6.2.1: Semax and Selank for Memory and Mood.. 73

    6.2.2: Dihexa and Cerebrolysin 75

  6.3: Long-Term Mental Health Protocols 76

CHAPTER 7: IMMUNE SYSTEM BOOST 78

  7.1: Peptides in Immune Health 78

  7.2: Best Peptide Combinations for Immunity 80

    7.2.1: Thymosin Alpha-1 for Immune Modulation 81

    7.2.2: LL-37 and KPV for Infection Defense 82

  7.3: Using Peptides for Preventative Health 84

    7.3.1: Seasonal Immune Protocols... 86

    7.3.2: Guidelines for Chronic Immune Support 87

CHAPTER 8: ANTI-AGING AND CELLULAR REGENERATION 90

  8.1: Peptides and Cellular Longevity 90

  8.2: Key Peptides for Skin Health 92

    8.2.1: Epithalon for Telomere Support 93

    8.2.2: GHK-Cu for Skin and Tissue Repair 94

  8.3: Protocols for Anti-Aging and Regeneration 95

PART 3: ADVANCED PEPTIDE PROTOCOLS 97

CHAPTER 9: PEPTIDE STACKS FOR SPECIFIC GOALS 98

  9.1: Fat Loss Stacks for Optimal Weight 98

  9.2: Brain Function Stacks for Clarity 99

  9.3: Sexual Health Peptide Stacks 101

  9.4: Skin and Cosmetic Peptide Stacks 102

CHAPTER 10: PEPTIDES, SUPPLEMENTS, AND LIFESTYLE 104

  10.1: Nutritional Support for Peptide Efficacy 104

  10.2: Supplement Pairings for Peptide Results 105

  10.3: Lifestyle Changes for Peptide Benefits 107

    10.3.1: Diet and Exercise Adjustments 108

    10.3.2: Stress Management for Peptide Performance..................................... 110

CHAPTER 11: MONITORING YOUR PROGRESS 112

  11.1: Biomarkers and Health Metrics 112

  11.2: Tools and Apps for Tracking Peptides 113

PART 4: PERSONALIZED FUTURE APPROACHES 115

CHAPTER 12: PERSONALIZED PEPTIDE PROTOCOLS 116

  12.1: Genetic and Personalized Peptide Use 116

    12.1.1: Utilizing DNA for Personalized Peptide Protocols 118

    12.1.2: Biomarker Monitoring for Optimized Outcomes 119

  12.2: Adapting Protocols to Lifestyle and Goals.. 120

CHAPTER 13: FUTURE TRENDS IN PEPTIDE THERAPY 122

  13.1: Cutting-Edge Research and Peptides 122

  13.2: Peptides in Personalized Medicine 124

  13.3: Ethical and Legal Considerations 125

CONCLUSION 128

  Summary of Key Protocols 128

  Maintaining Long-Term Health with Peptide Therapy 129

  Final Thoughts and Next Steps 130

Chapter 9: Peptide Stacks for Specific Goals 98
- 9.1: Fat Loss Stacks for Optimal Weight 98
- 9.2: Brain Function Stacks for Clarity 99
- 9.3: Sexual Health Peptide Stacks 101
- 9.4: Skin and Cosmetic Peptide Stacks 102

Chapter 10: Peptides, Supplements, and Lifestyle 104
- 10.1: Nutritional Support for Peptide Efficacy 104
- 10.2: Supplement Pairings for Peptide Results 105
- 10.3: Lifestyle Changes for Peptide Benefits 107
  - 10.3.1: Diet and Exercise Adjustments 108
  - 10.3.2: Stress Management for Peptide Performance 110

Chapter 11: Monitoring Your Progress 112
- 11.1: Biomarkers and Health Metrics 112
- 11.2: Tools and Apps for Tracking Peptides 113

Part 4: Personalized Future Approaches 115

Chapter 12: Personalized Peptide Protocols 116
- 12.1: Genetic and Personalized Peptide Use 116
  - 12.1.1: Utilizing DNA for Personalized Peptide Protocols 118
  - 12.1.2: Biomarker Monitoring for Optimized Outcomes 119
- 12.2: Adapting Protocols to Lifestyle and Goals 120

Chapter 13: Future Trends in Peptide Therapy 122
- 13.1: Cutting-Edge Research and Peptides 122
- 13.2: Peptides in Personalized Medicine 124
- 13.3: Ethical and Legal Considerations 125

Conclusion 128
- Summary of Key Protocols 128
- Maintaining Long-Term Health with Peptide Therapy 129
- Final Thoughts and Next Steps 130

Part 1:
Fundamentals of Peptide Science

Part 1:
Fundamentals of Peptide Science

Chapter 1:
Peptides 101

Peptides, short chains of amino acids linked by peptide bonds, are fundamental to understanding the intricate dance of cellular communication and function within the human body. These biological molecules, often overshadowed by their larger counterparts, proteins, hold the key to unlocking a myriad of health and wellness benefits, from enhancing muscle recovery to improving cognitive function. The distinction between peptides and proteins primarily lies in their size; peptides are smaller, making them quicker and easier for the body to absorb and utilize. This characteristic is particularly beneficial when discussing therapeutic applications, as their size allows for a more targeted approach to treatment, minimizing potential side effects.

The human body naturally produces peptides, playing critical roles in various physiological processes, including hormone production, immune response, and cell signaling. However, the body's ability to synthesize these peptides can diminish due to factors such as aging, stress, and environmental toxins, leading to a decrease in overall health and vitality. This is where synthetic peptides, created through biochemical research and development, come into play. Synthetic peptides mimic the structure and function of naturally occurring peptides, offering a promising avenue for health optimization.

Understanding how peptides work at the cellular level involves delving into the realm of receptors and signaling pathways. Peptides exert their effects by binding to specific receptors on the surface of cells, initiating a cascade of cellular responses that can alter physiological states. For example, certain peptides can signal the body to release growth hormone, which plays a vital role in muscle growth and repair, metabolism, and body composition. This receptor-ligand interaction is crucial for the therapeutic potential of peptides, allowing for precise modulation of biological processes.

The versatility of peptides as therapeutic agents is vast, with applications ranging from muscle building and fat loss to enhancing cognitive function and promoting deep, restorative sleep. The key to harnessing the full potential of peptide therapy lies in understanding the specific functions and mechanisms of action of different peptides. This knowledge enables the development of targeted peptide protocols that can address individual health concerns and goals, offering a personalized approach to health optimization.

As we delve deeper into the science of peptides, it's important to recognize the balance between natural and synthetic peptides. Each has its advantages and limitations, which will be explored further. The exploration of peptides as building blocks for health and vitality opens up a new frontier in the quest for ageless vitality and optimal well-being.

The exploration of natural versus synthetic peptides unveils a fascinating dichotomy in their origin and application. Natural peptides, those produced within the body or derived from natural sources, serve as essential components in numerous biological functions. Their roles in healing, immune function, and metabolic regulation underscore the body's inherent capacity for self-regulation and repair. However, the natural production of these peptides can wane due to the inevitable process of aging and lifestyle factors, leading to a gap in the body's ability to maintain optimal health.

Synthetic peptides, engineered in laboratories, are designed to fill this gap. By replicating the structure of natural peptides, scientists can create highly specific agents that target particular physiological pathways. The advantage of synthetic peptides lies in their precision and the ability to customize them for individual needs, offering a tailored approach to health optimization. Yet, this precision necessitates a deep understanding of peptide biology to avoid unintended consequences and ensure the safety and efficacy of peptide therapies.

The safety profile of peptides is generally favorable, especially when compared to more invasive treatments. However, the realm of synthetic peptides does present challenges in ensuring purity and proper usage. The potential for contamination or incorrect dosing underscores the importance of sourcing peptides from reputable suppliers and adhering to strict guidelines for their use. This is where the distinction between natural and synthetic peptides becomes most pronounced; while the body regulates natural peptide production, synthetic peptides require external oversight to ensure their safe application.

The therapeutic potential of peptides is vast, touching on nearly every aspect of health and wellness. From accelerating muscle recovery and fat loss to enhancing cognitive function and sleep quality, the applications of peptides are as diverse as the peptides themselves. The development of peptide protocols is an evolving science, requiring a nuanced understanding of each peptide's action and interaction within the body. This personalized approach to health optimization represents the future of peptide therapy, moving beyond one-size-fits-all solutions to offer targeted interventions that address individual health concerns and goals.

In conclusion, the exploration of peptides as therapeutic agents offers a promising avenue for enhancing human health and well-being. The distinction between natural and synthetic peptides, while significant, pales in comparison to the potential benefits that these molecules can offer. As research continues to unravel the complexities of peptide science, the potential for personalized, peptide-based therapies grows, heralding a new era in health optimization and disease prevention. The journey into the world of peptides is just beginning, with each discovery bringing us closer to unlocking the full potential of these remarkable molecules for improving human health.

1.1: WHAT ARE PEPTIDES?

Peptides are distinguished from proteins by their size, as they consist of fewer amino acids, generally less than 50. This smaller size contributes to their ability to penetrate the bloodstream more efficiently and interact with receptors on cell surfaces with high specificity. The structure of peptides includes a sequence of amino acids linked together by peptide bonds, a type of covalent chemical bond that forms when the carboxyl group of one amino acid reacts with the amino group of another. The sequence in which amino acids are arranged in a peptide chain is critical, as it determines the peptide's biological activity and function. Unlike proteins, which often fold into complex three-dimensional structures and can consist of multiple peptide chains, peptides usually adopt simpler structures. This simplicity allows them to be more easily synthesized and modified in the lab, providing a versatile tool for medical research and therapeutic development.

The biological activities of peptides are as diverse as their structures. Some peptides act as hormones, sending signals throughout the body to regulate physiological processes. Others have antimicrobial properties, playing a role in the body's immune response to infection. There are also peptides that influence how cells behave, promoting or inhibiting the growth of blood vessels, modulating immune responses, or directing cells to carry out specific tasks. The therapeutic potential of peptides is significant, given their ability to target specific cellular pathways with minimal side effects. This specificity stems from their ability to bind precisely to receptors on the surface of cells, initiating a cascade of cellular events without interfering with unrelated physiological processes.

The difference between natural and synthetic peptides is another critical aspect of peptide science. Natural peptides are produced within organisms and play various roles in normal physiological processes, from regulating metabolism to facilitating communication between cells. Synthetic peptides, on the other hand, are crafted in the laboratory. They offer the advantage of being customizable, allowing researchers to design peptides with specific sequences to target particular cellular processes or diseases. However, synthesizing peptides that perfectly mimic the structure and function of natural peptides can be challenging, requiring advanced technology and extensive knowledge of peptide chemistry.

In therapeutic applications, peptides are used to mimic or influence natural physiological processes. For example, peptides can be designed to stimulate the body's production of growth hormone, which can aid in muscle repair and growth, or to mimic the action of insulin, helping to regulate blood sugar levels. The development of peptide-based therapies involves rigorous testing to ensure that these molecules are effective and safe for human use. This includes determining the optimal dosage, understanding how the body metabolizes and excretes the peptide, and identifying any potential side effects.

The field of peptide research is rapidly advancing, with new peptides being discovered and synthesized regularly. These developments hold promise for the treatment of a wide range of conditions, from metabolic disorders and immune diseases to neurodegenerative disorders and cancer. As our understanding of peptides and their mechanisms of action expands, so too does the potential to harness these molecules for therapeutic purposes, offering hope for new treatments that are more targeted, effective, and with fewer side effects than traditional drugs. The exploration of peptides and their applications in medicine is a dynamic area of research, poised to make significant contributions to health and disease management.

1.2: PEPTIDES AT THE CELLULAR LEVEL

Once peptides bind to their respective cell surface receptors, a series of intricate cellular events is set into motion. This interaction triggers signal transduction pathways, a complex network of biochemical reactions that facilitate communication between the cell's external environment and its internal machinery. These pathways are crucial for translating extracellular signals into appropriate cellular responses, enabling the cell to adapt to various stimuli. The specificity of peptide-receptor interaction ensures that each signal is precisely targeted, minimizing unintended effects on other cellular processes. This high degree of specificity is what makes peptides so valuable in therapeutic applications, allowing for targeted intervention without disrupting the body's natural homeostasis.

Signal transduction often involves the activation of protein kinases, enzymes that modify other proteins by chemically adding phosphate groups—a process known as phosphorylation. This modification can alter a protein's activity, either activating or inhibiting its function, and is a common mechanism by which peptides exert their effects. For instance, the activation of certain kinases can lead to the transcription of genes associated with cell growth and repair, directly influencing processes such as muscle synthesis and recovery. Similarly, peptides that target metabolic pathways can enhance or suppress the expression of enzymes involved in fat metabolism, offering potential strategies for weight management and metabolic health.

Another critical aspect of peptide action at the cellular level is the regulation of ion channels. These channels control the flow of ions across the cell membrane, influencing cellular excitability, signaling, and homeostasis. Some peptides can modulate the activity of these channels, affecting processes such as muscle contraction, neurotransmitter release, and insulin secretion. This modulation can have profound effects on physiological functions, from improving muscle performance to regulating blood sugar levels.

Peptides can also influence cellular processes through their interaction with G-protein coupled receptors (GPCRs), a large family of receptors involved in a wide range of physiological responses. Upon binding to a GPCR, peptides can activate or inhibit various intracellular signaling cascades, leading to changes in cell behavior. This mechanism plays a significant role in the immune response, where peptides can modulate the activity of immune cells, enhancing the body's defense against pathogens or reducing inflammation.

The therapeutic potential of peptides is further enhanced by their ability to cross biological barriers, such as the blood-brain barrier (BBB). This capability allows certain peptides to exert their effects on the central nervous system, offering promising avenues for the treatment of neurological conditions. For example, peptides that can cross the BBB and bind to receptors in the brain have the potential to improve cognitive function, reduce symptoms of anxiety and depression, and support overall brain health.

Understanding the cellular mechanisms of peptides is essential for developing effective peptide-based therapies. By elucidating the pathways through which peptides exert their effects, researchers can design more targeted interventions, minimizing side effects and maximizing therapeutic benefits. This detailed knowledge of peptide action at the cellular level is what enables the development of personalized peptide protocols, tailored to the specific needs and health goals of each individual. The ability to customize peptide therapy based on a person's unique physiological profile represents a significant advancement in the field of health optimization, offering a more precise and effective approach to improving human health and well-being.

1.3: NATURAL VS. SYNTHETIC PEPTIDES

The distinction between **natural** and **synthetic peptides** is pivotal in understanding their role in therapeutic applications. Natural peptides, synthesized within living organisms, are integral to a myriad of biological processes. These peptides are involved in everything from hormone regulation to immune responses, showcasing the body's innate ability to maintain homeostasis and respond to internal and external stimuli. The production of natural peptides, however, can be influenced by a variety of factors including age, diet, stress levels, and environmental exposures. As such, the body's capacity to produce these peptides may decline, leading to imbalances or deficiencies that can affect overall health and well-being.

Synthetic peptides, on the other hand, are crafted through biochemical synthesis in the laboratory. This process allows for the creation of peptides with specific amino acid sequences tailored to target particular physiological pathways or processes. The advantage of synthetic peptides lies in their precision and the ability to be customized for individual therapeutic needs. Researchers can design synthetic peptides to mimic the action of natural peptides, potentially compensating for deficiencies or imbalances within the body. Moreover, synthetic peptides can be engineered to enhance stability, increase bioavailability, or reduce potential side effects, making them highly effective tools in drug development and therapeutic interventions.

One of the primary advantages of **natural peptides** is their compatibility with the body's biological systems. Being naturally occurring, these peptides are generally well-tolerated with minimal risk of adverse reactions. They are recognized by the body's receptors and can effectively engage in physiological processes without the need for modification. However, the extraction and purification of natural peptides can be challenging and costly. Additionally, the availability of natural peptides is limited by the source from which they are derived, which may impact scalability and accessibility for therapeutic use.

Synthetic peptides offer a solution to these limitations by providing a scalable and customizable approach to peptide therapy. The ability to design peptides with specific sequences allows for targeted therapeutic applications, potentially increasing efficacy and reducing side effects. Synthetic peptides can also be produced in large quantities, ensuring consistency and accessibility for research and clinical use. However, the synthesis of peptides requires sophisticated technology and expertise, which can be resource-intensive. Furthermore, the introduction of synthetic peptides into the body may pose a risk of immune reactions or other adverse effects if not carefully designed and tested.

The choice between natural and synthetic peptides often depends on the specific application and desired outcome. For instance, in cases where a natural peptide is known to play a critical role in a physiological process but is deficient or imbalanced in the body, a synthetic peptide designed to mimic the natural molecule can be an effective therapeutic strategy. On the other hand, leveraging the natural production of peptides through lifestyle interventions, such as diet and exercise, may be preferred for general health optimization and prevention strategies.

In the realm of peptide therapy, both natural and synthetic peptides hold significant promise. The ongoing research and development in peptide science continue to expand our understanding of these molecules and their potential applications. As we advance, the ability to harness the benefits of both natural and synthetic peptides will be crucial in developing personalized and effective therapeutic strategies. The integration of peptide therapies into clinical practice offers a new frontier in medicine, with the potential to address unmet medical needs and improve patient outcomes across a wide range of conditions.

1.4: Peptides for Health Optimization

Peptides play a pivotal role in health optimization by targeting four key areas: regeneration, immune support, anti-aging, and energy enhancement. The body's natural ability to heal and regenerate is fundamental to maintaining optimal health. Peptides such as BPC-157 and TB-500 have shown remarkable potential in accelerating wound healing and tissue regeneration. BPC-157, often referred to as the "body protection compound," has been observed to significantly enhance the healing of tendons, muscles, nervous system, and even the gut. TB-500, on the other hand, plays a crucial role in the regeneration of blood vessels, promoting improved circulation and faster recovery from injuries. These peptides work by upregulating the expression of specific genes involved in the healing process, thereby facilitating more efficient tissue repair and regeneration.

Immune support is another critical benefit of peptide therapy. Thymosin Alpha-1, for example, has been extensively studied for its ability to modulate the immune system. It enhances the body's ability to fight off infections and diseases by increasing the activity of T-cells and dendritic cells, which are essential components of the immune response. This peptide can be particularly beneficial for individuals with weakened immune systems, offering a boost that helps the body defend against pathogens more effectively.

When it comes to anti-aging, peptides like Epithalon and GHK-Cu have shown promising results in slowing down the aging process and improving skin health. Epithalon works by lengthening telomeres, the protective caps at the ends of chromosomes, which naturally shorten as we age. By maintaining telomere length, Epithalon can potentially extend cell life and promote longevity. GHK-Cu, a copper peptide, has been found to improve skin elasticity, reduce fine lines and wrinkles, and stimulate collagen production, leading to a more youthful appearance. These anti-aging peptides not only contribute to better skin health but also support overall cellular regeneration and longevity.

Energy enhancement is another significant benefit of peptide therapy. Peptides like MOTS-c and AOD-9604 have been shown to improve metabolic health and energy levels. MOTS-c, in particular, activates metabolic pathways that improve insulin sensitivity, enhance glucose regulation, and promote fat burning, leading to increased energy and improved metabolic health. AOD-9604, originally developed as an anti-obesity drug, has been found to stimulate lipolysis (the breakdown of fats) and inhibit lipogenesis, the process by which fats are synthesized and stored in the body. By enhancing fat metabolism, AOD-9604 can help improve energy levels and support weight management.

Chapter 1:
Peptides 101

Incorporating peptide therapy into a health optimization plan requires a personalized approach, taking into account individual health goals, conditions, and potential interactions with other treatments. It's essential to work with a healthcare provider experienced in peptide therapy to determine the most appropriate peptides and dosages for your specific needs. Additionally, sourcing peptides from reputable suppliers is crucial to ensure purity and efficacy. By carefully selecting and properly administering peptide therapies, individuals can harness the significant benefits peptides offer for regeneration, immune support, anti-aging, and energy enhancement, contributing to overall health optimization and well-being.

Chapter 2: Quality, Safety, and Sourcing Essentials

2.1: Ensuring Peptide Quality

When considering the purchase and use of peptides for health optimization, ensuring the quality of these compounds is paramount. The first step in this process involves identifying key parameters that signify peptide quality, such as **purity** and **authenticity**. Purity refers to the percentage of the peptide that is the actual intended compound, free from contaminants or other adulterants. Authenticity ensures that the peptide matches its specification in terms of structure and function. Both of these factors are critical, as impurities can not only reduce the effectiveness of the peptide but also pose significant health risks.

Laboratory testing plays a crucial role in verifying these parameters. High-performance liquid chromatography (HPLC) is a common method used to assess peptide purity. This technique separates the components of a mixture to identify and quantify the target peptide. Mass spectrometry (MS) is another sophisticated technique that provides information about the molecular weight and structure of the peptide, confirming its identity and purity. Certificates of Analysis (CoA) from reputable laboratories offer a detailed breakdown of these test results, providing assurance about the peptide's quality.

Another aspect to consider is the source of the peptides. Reputable suppliers will have stringent quality control processes in place, from synthesis to packaging and shipping. They should be transparent about their manufacturing processes and willing to provide documentation to back up their quality claims. It's also advisable to research suppliers, looking for reviews or testimonials from other users, which can offer insights into the reliability and quality of their products.

Peptide stability and shelf-life are additional factors that impact quality. Peptides can degrade over time or when exposed to certain conditions, such as high temperatures or moisture. Suppliers should provide information on the proper storage and handling of peptides to maintain their integrity until use. This often includes recommendations for refrigeration or freezing and guidelines on reconstituting lyophilized (freeze-dried) peptides.

In summary, ensuring peptide quality requires attention to purity and authenticity, backed by rigorous laboratory testing. Sourcing from reputable suppliers, understanding stability and storage requirements, and reviewing certificates of analysis are all essential steps in securing safe and effective peptides for health optimization. By prioritizing these quality measures, users can confidently incorporate peptides into their wellness routines, maximizing the potential benefits while minimizing risks.

Chapter 2:
Quality, Safety, and Sourcing Essentials

2.2: PEPTIDE PURITY AND POTENCY

Understanding the critical aspects of peptide purity and potency is essential for ensuring safety in their use. The purity of a peptide refers to the proportion of the peptide that is the intended compound, devoid of any contaminants or other substances. Potency, on the other hand, measures the biological activity of the peptide, indicating how effective it is in achieving its intended effects at a given dose. Both these factors are paramount in determining the safety and efficacy of peptide protocols for health optimization.

Contaminants present in peptides can include other peptides, bacterial endotoxins, or heavy metals, which can arise from substandard manufacturing processes. The presence of these contaminants can lead to adverse reactions, ranging from mild allergic responses to severe health complications. For instance, impurities in peptides can trigger immune responses, leading to inflammation or other immune-mediated reactions. Similarly, peptides with a lower than stated potency may require higher doses to achieve the desired effect, potentially increasing the risk of side effects.

To mitigate these risks, it is crucial to source peptides from reputable suppliers who adhere to stringent manufacturing standards. Look for suppliers that utilize Good Manufacturing Practices (GMP), as this indicates a commitment to quality and safety. GMP covers all aspects of production, from the starting materials, premises, and equipment to the training and personal hygiene of staff. Detailed written procedures are essential for each process that could affect the quality of the finished product. Systems must be in place to provide documented proof that correct procedures are consistently followed at each step in the manufacturing process - every time a product is made.

Moreover, third-party laboratory testing becomes an indispensable tool in verifying the purity and potency of peptides. Independent labs can provide unbiased reports on the composition and strength of the peptides, offering an additional layer of assurance. When evaluating these reports, pay close attention to the methods used for testing. High-performance liquid chromatography (HPLC) and mass spectrometry (MS) are gold standards for determining purity, while bioassays or enzyme-linked immunosorbent assays (ELISA) are commonly used to assess potency.

In addition to sourcing and testing, proper storage and handling of peptides are vital to maintain their integrity. Peptides can be sensitive to environmental conditions, such as light, temperature, and moisture, which can degrade their quality over time. Adhering to the manufacturer's guidelines for storage, such as keeping peptides refrigerated or frozen and protected from light, can help preserve their potency and reduce the risk of degradation.

In conclusion, the safety and effectiveness of peptide protocols hinge on the purity and potency of the peptides used. By prioritizing these factors and adhering to best practices in sourcing, testing, and storage, individuals can maximize the health benefits of peptides while minimizing potential risks. This approach ensures that the therapeutic potential of peptides is fully realized, supporting health optimization goals with confidence.

2.3: PEPTIDES LEGALITY AND REGULATIONS

Navigating the complex landscape of **peptide legality and regulatory considerations** in the United States requires a thorough understanding of the **Food and Drug Administration (FDA)** guidelines and the **Controlled Substances Act**. The FDA regulates peptides under the category of drugs or biologics depending on their intended use, mechanism of action, and method of administration. This classification significantly impacts the approval process, labeling, marketing, and distribution of peptide products. For peptides considered drugs, obtaining FDA approval involves demonstrating safety and efficacy through rigorous clinical trials. Biologics, including some peptides, undergo a similar but distinct approval process that emphasizes biological source control and manufacturing processes.

The **Controlled Substances Act** further complicates the legal landscape for peptides by classifying certain peptides as controlled substances based on their potential for abuse and dependence. This classification restricts their manufacture, distribution, and possession, requiring researchers, healthcare providers, and consumers to navigate a maze of regulatory requirements to ensure compliance. For instance, peptides with performance-enhancing properties might be closely monitored or restricted, emphasizing the importance of understanding federal and state laws before purchasing or using peptides.

Moreover, the **Dietary Supplement Health and Education Act (DSHEA)** of 1994 provides another layer of regulation by differentiating dietary supplements from drugs and biologics. Under DSHEA, peptides marketed as dietary supplements must not claim to treat, diagnose, prevent, or cure diseases, a restriction that significantly limits the marketing of peptides for health optimization purposes. Manufacturers and distributors must ensure that their products comply with DSHEA guidelines, including label claims, manufacturing practices, and adverse event reporting.

Chapter 2:
Quality, Safety, and Sourcing Essentials

State laws may also impose additional restrictions on the sale and use of peptides, with some states enacting stricter controls than federal regulations. This patchwork of state-specific regulations necessitates a careful review of local laws to avoid legal pitfalls, particularly for businesses operating in multiple states or for individuals ordering peptides online from out-of-state suppliers.

For consumers looking to purchase peptides legally and safely, several practical tips can help navigate these regulatory waters. First, verify the regulatory status of the peptide of interest, ensuring it is legal for purchase and use in your state. Second, purchase peptides only from reputable suppliers who comply with FDA and DSHEA regulations, providing transparent product information, including sources, purity, and intended use. Third, consult healthcare providers or legal experts when in doubt about the legality or safety of a peptide product, especially if intended for therapeutic use.

In the realm of **peptide research and development**, staying abreast of regulatory changes is crucial. The FDA and other regulatory bodies periodically update guidelines and approval processes, reflecting new scientific evidence and safety concerns. Researchers and companies involved in peptide development should engage with regulatory agencies early in the development process, seeking guidance and ensuring compliance with the latest standards.

Finally, the evolving legal landscape underscores the importance of advocacy and education within the peptide community. By staying informed and engaged, researchers, healthcare providers, and consumers can contribute to a regulatory environment that balances safety and innovation, allowing the therapeutic potential of peptides to be fully realized within a legal framework. This proactive approach not only ensures compliance but also fosters the responsible use and development of peptides for health optimization.

2.3.1: Understanding US Peptide Regulations

The regulatory framework governing peptides in the United States is multifaceted, with the Food and Drug Administration (FDA) playing a pivotal role in ensuring that peptides used for therapeutic purposes meet stringent safety and efficacy standards. The FDA categorizes peptides as either drugs or biologics based on their intended use, composition, and method of administration. This classification is crucial as it dictates the pathway for regulatory approval, which involves comprehensive clinical trials to demonstrate safety and effectiveness for drugs, and a focus on source control and manufacturing processes for biologics. The complexity of these regulations underscores the importance of compliance for manufacturers and distributors, who must navigate these pathways to bring their peptide products to market.

Under the Controlled Substances Act, certain peptides are classified as controlled substances, which imposes additional layers of regulation. This classification aims to prevent misuse and abuse, particularly for peptides that have performance-enhancing properties. Manufacturers, healthcare providers, and consumers must be vigilant in adhering to these regulations, which include restrictions on manufacture, distribution, and possession. It is essential to understand both federal and state laws, as state regulations may introduce stricter controls beyond federal requirements. This dual layer of regulation necessitates a careful, informed approach to ensure legal compliance and avoid potential legal repercussions.

The Dietary Supplement Health and Education Act (DSHEA) of 1994 further complicates the regulatory landscape by distinguishing between dietary supplements and drugs or biologics. Peptides marketed as dietary supplements must not make claims to treat, diagnose, prevent, or cure diseases. Compliance with DSHEA guidelines is critical for manufacturers and distributors to avoid penalties and ensure that their products can be legally marketed in the United States. This includes adherence to specific labeling requirements, manufacturing practices, and adverse event reporting protocols.

For consumers seeking to use peptides for health optimization, it is paramount to source these compounds from reputable suppliers who adhere to FDA and DSHEA regulations. This includes verifying the legal status of peptides, ensuring that the supplier provides transparent product information, and consulting healthcare providers or legal experts when necessary. Purchasing peptides from compliant sources not only supports legal and safe use but also contributes to the broader goal of ensuring that peptide products available in the market are of high quality and efficacy.

In the realm of peptide research and development, staying informed about regulatory changes is critical. The FDA and other regulatory bodies periodically update their guidelines and approval processes to reflect new scientific insights and safety considerations. Engagement with regulatory agencies from the early stages of peptide development can facilitate compliance and streamline the approval process. This proactive approach benefits researchers and companies by providing clarity on regulatory expectations and helping to avoid costly delays or adjustments.

The peptide community, including researchers, healthcare providers, and consumers, plays a vital role in navigating the regulatory environment. By staying informed, engaged, and compliant, the community can support the responsible development and use of peptides. This collective effort is essential for harnessing the therapeutic potential of peptides within a legal and regulatory framework that prioritizes safety and innovation. As the legal landscape evolves, continuous education and advocacy will be key to adapting to new regulations and ensuring that peptides continue to offer health optimization benefits within a framework of legal and regulatory compliance.

2.3.2: Global Legal Landscape for Peptide Use

The global legal landscape for peptide use presents a complex patchwork of regulations that vary significantly from one country to another, reflecting diverse approaches to healthcare, pharmaceutical regulation, and biotechnology. This variability necessitates a careful, country-specific analysis for individuals and entities involved in the peptide field, whether for personal use, research, or commercial purposes. Understanding these differences is crucial for navigating international laws and ensuring legal compliance across borders.

In the European Union (EU), for example, peptides are regulated under a framework that emphasizes consumer safety and product efficacy. The European Medicines Agency (EMA) plays a central role in the evaluation and approval of new peptide-based therapies, requiring a rigorous demonstration of safety, quality, and efficacy before they can be marketed. The EU's regulatory approach also includes strict guidelines on manufacturing practices, labeling, and advertising, aiming to protect consumers from misleading claims and unsafe products. Additionally, the EU's classification of peptides can affect their availability and use, with some peptides considered medicinal products and others as research compounds, each subject to different regulatory pathways and controls.

Contrastingly, in countries like Japan, the regulatory environment for peptides is shaped by the Pharmaceuticals and Medical Devices Agency (PMDA), which oversees the approval and monitoring of pharmaceuticals, including peptide therapies. Japan's regulatory framework is known for its rigorous approval process, which includes extensive clinical trials and safety evaluations. However, Japan has also been at the forefront of incorporating regenerative medicine into its healthcare system, which has implications for peptide therapies and research. This progressive stance reflects a balance between stringent regulatory oversight and support for innovative treatments.

In emerging markets, such as Brazil, India, and China, the regulatory landscape for peptides is evolving rapidly. These countries are working to strengthen their regulatory frameworks to support the growth of their biopharmaceutical sectors, protect public health, and meet international standards. This includes establishing clear guidelines for the approval, manufacture, and sale of peptide products. However, regulatory capacity and enforcement can vary, presenting challenges for ensuring product quality and safety. As these markets continue to develop their regulatory infrastructures, they offer significant opportunities for the peptide industry, alongside the need for vigilance in compliance and quality assurance.

Navigating the global legal landscape for peptide use requires a multifaceted strategy. For individuals seeking to use peptides for personal health optimization, it is essential to research the legal status of specific peptides in their country, including any restrictions on importation or use. For researchers and companies, understanding the regulatory requirements for clinical trials, marketing approval, and post-market surveillance in different jurisdictions is key to global operations. This may involve engaging with regulatory consultants or legal experts who specialize in pharmaceutical and biotech law in target markets.

Moreover, the international trade of peptides, whether for research or commercial purposes, must comply with export and import regulations, including customs declarations, permits, and inspections. This is particularly relevant for peptides that are classified under controlled substances or dual-use regulations, which are subject to additional scrutiny and control to prevent misuse.

The global legal landscape for peptide use underscores the importance of staying informed and adaptable. Regulatory environments are continually changing, influenced by advances in science, public health priorities, and international harmonization efforts. For the peptide community, this dynamic landscape presents both challenges and opportunities. By staying abreast of legal developments and engaging proactively with regulatory authorities, individuals and organizations can navigate the complexities of international peptide use with confidence, ensuring legal compliance and contributing to the advancement of peptide science and therapy.

2.3.3: Safe and Legal Peptide Purchase Tips

When embarking on the purchase of peptides, the primary concern should be to ensure that the transaction is both safe and legal. This necessitates a meticulous approach to verifying the credibility and compliance of peptide suppliers. The first step in this process involves conducting thorough research on potential suppliers.

Chapter 2:
Quality, Safety, and Sourcing Essentials

This research should extend beyond the suppliers' websites to include reviews and feedback from previous customers, which can often be found on forums, social media platforms, and independent review sites. Such reviews can provide invaluable insights into the quality of the peptides supplied, as well as the supplier's customer service and reliability.

Another critical step is to request and review the **Certificates of Analysis (CoA)** for any peptides you are considering purchasing. A CoA is a document issued by a third-party laboratory that confirms the peptide's purity and composition. This document should detail the specific tests that were conducted, such as **High-Performance Liquid Chromatography (HPLC)** and **Mass Spectrometry (MS)**, and the results of these tests. By examining the CoA, you can ensure that the peptide meets the necessary purity standards and is free from harmful contaminants.

It is also advisable to verify the supplier's compliance with **Good Manufacturing Practices (GMP)**. GMP certification indicates that the supplier adheres to stringent quality control standards throughout the manufacturing process, from the sourcing of raw materials to the final packaging and distribution of the peptides. Suppliers who are GMP certified are more likely to provide peptides that are safe, pure, and effective.

Engaging directly with the supplier can provide further assurance of their credibility and the quality of their products. Reputable suppliers should be transparent about their manufacturing processes and willing to answer any questions you may have regarding their peptides' origin, purity, and safety. They should also be knowledgeable about the legal regulations governing the sale and use of peptides in your jurisdiction and able to provide guidance on legal compliance.

When it comes to the legal aspects of purchasing peptides, it is essential to be aware of the regulations that apply in your country or state. In the United States, for example, the legal status of peptides can vary depending on their intended use and whether they are classified as research chemicals or intended for human consumption. Purchasing peptides labeled for research purposes may circumvent certain regulatory restrictions, but it is crucial to understand that such peptides are not intended for human use and doing so could pose significant health risks.

For those looking to use peptides therapeutically, consulting with a healthcare professional is an important step. A healthcare provider can offer guidance on the appropriate use of peptides, potential health benefits and risks, and legal considerations. In some cases, peptides may be prescribed by a doctor, which can ensure that their use is both safe and compliant with legal regulations.

In summary, ensuring the safe and legal purchase of peptides requires a proactive approach to supplier research, a thorough examination of product quality and compliance, and an understanding of the legal landscape. By taking these steps, individuals can confidently navigate the complexities of the peptide market, securing high-quality peptides that meet their health and wellness goals while adhering to legal standards.

Chapter 3: Practical Guidelines for Peptides

3.1: Determining the Right Dosage

When determining the **optimal dosage** for peptide use, it's crucial to consider both your **physical condition** and **personal health goals**. The process involves a nuanced understanding of how peptides interact with your body at a cellular level and the specific outcomes you wish to achieve, whether it's enhancing muscle recovery, improving sleep quality, achieving sharper focus, or supporting overall vitality. The

first step in this process is to **assess your current health status** and any specific conditions that might influence how your body responds to peptides. This assessment could involve consulting with a healthcare professional to understand any underlying health issues that could affect peptide efficacy or safety.

Secondly, it's important to **clearly define your health goals**. Are you looking to peptides primarily for their potential to support muscle growth and recovery, or are you more interested in their anti-aging and cognitive benefits? The specific peptides and their dosages will vary significantly based on the outcomes you're aiming for. For instance, peptides like **CJC-1295** and **Ipamorelin** might be more suited for those looking to enhance growth hormone production for muscle development, whereas **Semax** and **Selank** could be better options for those focusing on cognitive health improvements.

Thirdly, understanding the **mechanism of action** of each peptide is key to determining the right dosage. Peptides can act through various pathways, such as mimicking natural growth hormone-releasing hormones or directly stimulating the production of certain hormones or proteins. The dosage will depend on the peptide's potency and the desired speed and intensity of the results. It's also important to consider the **half-life** of the peptide, as this will dictate the frequency of dosing to maintain its therapeutic effects.

Fourth, the method of **administration** plays a critical role in dosage determination. Peptides can be administered through injections, which is the most common and effective method, or orally, topically, and nasally, depending on the peptide's formulation. The bioavailability of the peptide, or the proportion which enters the circulation when introduced into the body and so is able to have an active effect, varies significantly with the method of administration, influencing the dosage needed to achieve the desired effect.

Fifth, it's essential to **start with a lower dosage** and gradually increase it based on your body's response. This approach, often referred to as **titration**, helps minimize potential side effects and allows you to find the minimum effective dose that provides the desired benefits. For example, when using peptides for muscle recovery, starting with a lower dose and observing the effects on muscle soreness and recovery time can guide adjustments to the dosage.

Lastly, ongoing **monitoring and adjustment** are crucial. As your body adapts to the peptide or as your health goals evolve, adjustments to the dosage may be necessary. Regular consultations with a healthcare provider, along with monitoring of relevant biomarkers, can provide insights into how well the peptide protocol is working and whether any changes are needed.

Chapter 3:
Practical Guidelines for Peptides

In summary, determining the right peptide dosage is a personalized process that requires careful consideration of your health status, goals, the specific peptides being used, their mechanisms of action, and the method of administration. Starting with a conservative dose and adjusting based on your body's response and ongoing monitoring will help optimize the benefits of peptide use while minimizing risks.

3.1.1: Calculating Dosages by Weight and Goals

Calculating dosages based on **body weight and specific goals** requires a meticulous approach, as the effectiveness and safety of peptide protocols hinge on these personalized factors. The process begins with an accurate measurement of body weight, which serves as a foundational metric for dosage calculations. For peptides that are dosed according to body weight, the general guideline is to use micrograms (mcg) or milligrams (mg) per kilogram (kg) of body weight. This calculation ensures that the dosage is proportional to the individual's mass, optimizing the potential for desired outcomes while minimizing the risk of adverse effects.

For instance, if a peptide protocol suggests a dosage of 1 mcg/kg for enhancing muscle recovery, an individual weighing 70 kg would calculate their dose as follows: 1 mcg multiplied by 70 kg equals a 70 mcg dose. This method is straightforward yet critical for tailoring peptide use to individual needs and goals. It's also important to note that these calculations serve as a starting point. Individuals may require adjustments based on their response to the peptide, necessitating a period of observation and fine-tuning.

Specific goals further refine dosage calculations. For **muscle growth**, higher dosages within the safe range may be more effective, leveraging the anabolic properties of certain peptides. Conversely, goals centered on **anti-aging** or **cognitive enhancement** might benefit from lower, more frequent doses, focusing on sustained activation of regenerative pathways without overstimulation. The distinction between these objectives underscores the importance of understanding the biological mechanisms at play and how they align with personal health aspirations.

Moreover, the **half-life** of peptides must be considered in these calculations. Peptides with a shorter half-life may require more frequent dosing to maintain their therapeutic window, whereas those with a longer half-life might be administered less often. This aspect of peptide pharmacokinetics plays a crucial role in achieving consistent and effective results, ensuring that the body receives a steady signal for repair, growth, or cognitive support.

Incorporating these factors into a cohesive dosage strategy also involves acknowledging the dynamic nature of individual health and fitness journeys. As body weight fluctuates and goals evolve, so too will the optimal peptide dosage. Regular reassessment of these parameters ensures that peptide protocols remain aligned with current needs, maximizing benefits while safeguarding health.

It's also imperative to source peptides from reputable suppliers and consult with healthcare professionals experienced in peptide therapy to validate dosage calculations and protocols. This collaborative approach facilitates a customized peptide regimen that is both effective and aligned with best practices in health and wellness.

By adhering to these principles of dosage calculation based on body weight and specific goals, individuals can harness the full potential of peptides to support their health optimization efforts. Whether the aim is to enhance physical performance, accelerate recovery, improve cognitive function, or promote longevity, the precision in dosing is a cornerstone of success in biohacking with peptides.

3.1.2: Dosage Guidelines for Beginners vs. Advanced

When transitioning from beginner to advanced peptide protocols, the approach to dosage must be carefully adjusted to reflect the user's growing experience and physiological adaptation to peptide therapy. For beginners, the primary focus should be on establishing a baseline tolerance and understanding of how one's body reacts to peptide administration. This phase is critical for ensuring safety and minimizing potential adverse reactions. Beginners should start with the lowest recommended dose of any peptide protocol. This conservative approach allows the body to adjust gradually, reducing the risk of side effects and enabling the user to monitor and evaluate the peptide's effects on their body. It is advisable for beginners to maintain this initial dosage for a minimum period, often several weeks, before considering any adjustments. This period of observation is crucial for assessing the body's response and determining whether the desired outcomes are being achieved without negative consequences.

As users transition from beginner to advanced stages, they have the opportunity to explore higher dosages or more complex peptide stacks, always within the safety parameters established by clinical research and medical advice. Advanced users typically have a well-established tolerance to peptides and a deeper understanding of how different peptides affect their body and health goals. At this stage, users may consider incrementally increasing their dosage to enhance the desired effects, whether

Chapter 3:
Practical Guidelines for Peptides

for muscle growth, fat loss, cognitive function, or overall vitality. However, this increase should be done juditiously, with careful monitoring of any changes in body response or the emergence of side effects. It is also at this advanced stage that individuals can experiment with peptide stacking, where multiple peptides are used in combination to target specific health goals more effectively. Peptide stacking should be approached with a strategic mindset, understanding the synergistic potential of certain peptides when used together and the importance of maintaining safe dosage levels across the stack.

For both beginners and advanced users, it is paramount to maintain an ongoing dialogue with healthcare professionals experienced in peptide therapy. This ensures that any adjustments to dosage or protocol are made with a comprehensive understanding of the individual's health status, goals, and the scientific evidence supporting peptide use. Regular blood work and other relevant health metrics can provide objective data to guide these adjustments, ensuring that the peptide protocol remains both safe and effective over time.

Moreover, advanced users must remain vigilant about the quality and sourcing of their peptides. As dosages increase or as more complex stacks are introduced, the importance of using high-purity, accurately dosed peptides becomes even more critical. This underscores the necessity of sourcing peptides from reputable suppliers who provide third-party testing results, confirming the purity and potency of their products.

In conclusion, transitioning from beginner to advanced peptide use involves a nuanced understanding of dosage adjustments, the physiological responses to peptides, and the strategic use of peptide stacking to achieve specific health goals. By adhering to a cautious and informed approach, users can safely explore the benefits of peptide therapy, optimizing their health and well-being while minimizing risks.

3.2: SAFE INJECTION PRACTICES AND STORAGE TIPS

Moving forward, it's imperative to discuss the **safe injection practices** essential for minimizing risks and ensuring the efficacy of peptide therapies. Injection techniques are a cornerstone of peptide administration, with subcutaneous (SC) and intramuscular (IM) injections being the most common methods. Each technique has its specific protocols, which, when followed meticulously, can significantly reduce the risk of complications such as infections or tissue damage.

Subcutaneous injections are performed by inserting a short needle into the fatty tissue just beneath the skin. This method is widely used for peptides that require slow, steady absorption into the bloodstream. The key to a successful SC injection lies in selecting the proper injection site, typically the abdomen or thigh, ensuring the area is clean and disinfected. Pinching the skin can help in creating a more defined area for injection, making the process smoother and less painful.

Intramuscular injections, on the other hand, involve delivering the peptide directly into the muscle. This technique allows the peptide to be absorbed more quickly compared to the SC route. Common IM injection sites include the deltoid, vastus lateralis, and ventrogluteal muscles. As with SC injections, cleanliness cannot be overstressed; the injection site must be thoroughly cleaned with an alcohol swab prior to injection. Additionally, rotating sites between injections can help prevent muscle soreness and tissue damage.

Regardless of the method chosen, using the correct needle size is crucial. A needle that is too long can cause unnecessary pain or inject the peptide too deeply, while one that is too short may not deliver the peptide effectively. Generally, a 1/2 to 5/8 inch, 25 to 30 gauge needle is suitable for SC injections, and a 1 to 1.5 inch, 22 to 25 gauge needle for IM injections.

Storage and handling guidelines are equally important for maintaining the potency and purity of peptides. Peptides often come in lyophilized (freeze-dried) form and require reconstitution with bacteriostatic water or sterile saline before use. Once reconstituted, peptides must be stored in the refrigerator to prevent degradation. The peptide solution should be gently mixed, avoiding vigorous shaking that can damage the peptide structure. Proper labeling of reconstituted peptides with the date and time of mixing can help track their usability period, as most peptides are stable for weeks to months when stored correctly.

Avoiding contamination is a critical aspect of peptide handling. Always use sterile syringes and needles for each injection and dispose of them properly in a sharps container. Never share needles or syringes, as this can lead to serious infections. If a peptide solution appears cloudy, discolored, or contains particles, it should not be used, as these are signs of contamination or degradation.

In conclusion, adhering to safe injection practices and proper storage guidelines is paramount for anyone utilizing peptide therapies. These practices ensure the maximum efficacy of the peptides while minimizing potential risks associated with their administration. By following these guidelines, users can confidently incorporate peptides into their health optimization routines, unlocking the myriad benefits these powerful molecules have to offer.

3.2.1: Peptide Storage and Handling Guidelines

Upon reconstitution of peptides with bacteriostatic water or sterile saline, it is crucial to understand that their stability and efficacy are now subject to degradation over time if not stored correctly. The refrigerator becomes an essential tool in preserving the integrity of these substances, with a general guideline suggesting storage at temperatures between 2°C to 8°C (35°F to 46°F). This temperature range is optimal for slowing down the degradation process, thereby extending the shelf life of peptides. However, it's important to note that not all peptides have the same storage requirements post-reconstitution; some may have a significantly shorter usable life even under refrigeration. Therefore, consulting the peptide's specific storage instructions provided by the manufacturer is a step that cannot be overlooked.

For peptides that are supplied in lyophilized form and have yet to be reconstituted, a cool, dry place away from direct sunlight is the preferred storage condition. These peptides, in their freeze-dried state, are relatively stable and can withstand room temperature for extended periods. However, to maximize their shelf life, storing them in a refrigerator—even in their lyophilized state—is advisable, especially if they will not be used immediately.

The process of reconstitution itself requires careful attention to detail to avoid compromising the peptide's purity and potency. Using sterile techniques to draw the reconstitution fluid and inject it into the peptide vial is paramount. The introduction of the fluid should be gentle, allowing it to naturally flow down the side of the vial rather than directly onto the lyophilized powder. This technique minimizes the risk of damaging the peptide structure. Once added, the vial should be swirled gently rather than shaken vigorously. Shaking can lead to the formation of bubbles, which can denature some peptides, rendering them less effective or entirely ineffective.

After reconstitution, peptides in solution form should be used within a timeframe that ensures their potency. This period can vary widely among different peptides, with some remaining stable for only a few days, while others may last for weeks. Again, the manufacturer's guidelines can provide specific insights into the expected stability of each peptide. For peptides with a shorter lifespan post-reconstitution, preparing smaller quantities to minimize waste is a strategy that can be employed.

Labeling of reconstituted peptides is another critical practice that ensures safe and effective use. Each vial should be clearly labeled with the name of the peptide, the concentration of the solution, the date of reconstitution, and the expiration date based on the peptide's stability. This practice not only helps in maintaining an organized storage system but also reduces the risk of using a peptide past its effective date or confusing one peptide for another.

The handling of peptides, both in their lyophilized and reconstituted forms, should always be done with clean, sanitized hands, and preferably in a clean area free from potential contaminants. The use of gloves is recommended to prevent the transfer of oils and bacteria from the skin to the peptide containers. Additionally, all equipment used in the reconstitution process, such as syringes and needles, should be sterile and for single use only to prevent contamination.

In managing the storage and handling of peptides, the overarching goal is to preserve their structural integrity and biological activity. This careful management ensures that when administered, the peptides can perform their intended function without compromise. Whether for research purposes or personal use under the guidance of a healthcare professional, adhering to these guidelines is essential for achieving the desired outcomes from peptide therapies.

3.2.2: Subcutaneous vs. Intramuscular Injections

Choosing between subcutaneous and intramuscular injections is a decision that hinges on the specific peptides being used, their intended effects, and individual comfort levels. Subcutaneous injections, being less invasive, are often preferred for daily or frequent dosing schedules. The technique involves a gentle insertion of a short, thin needle into the fatty layer beneath the skin, a method that facilitates slow and steady absorption of peptides into the bloodstream. This method is particularly suited for

Chapter 3:
Practical Guidelines for Peptides

peptides that require gradual systemic distribution to exert their effects over time. The abdomen, with its ample fatty tissue, provides an ideal site for such injections, offering not only ease of access but also the benefit of rotating injection sites to minimize tissue irritation.

In contrast, intramuscular injections deliver peptides directly into the muscle, allowing for a faster absorption rate. This method is advantageous for peptides that necessitate rapid uptake to achieve desired outcomes, such as those used for muscle recovery and growth. The choice of muscle group for injection should consider both the volume of the peptide solution and the frequency of injections; the deltoid, for example, is suitable for smaller volumes, while the vastus lateralis muscle of the thigh can accommodate larger quantities. It's crucial to vary the injection sites within the chosen muscle to prevent the formation of scar tissue, which can hinder the absorption of peptides.

The technique for both injection types requires precision and care. For subcutaneous injections, creating a small fold of skin between the thumb and forefinger lifts the fatty tissue away from the underlying muscle, providing a cushion into which the needle can be inserted safely. This method reduces the risk of injecting into muscle tissue, which is not the intended target for this type of injection. Intramuscular injections, meanwhile, demand an understanding of the muscle's anatomy to avoid nerves and blood vessels. The use of a longer needle enables the peptide to penetrate deeply into the muscle tissue, ensuring that the substance is deposited where it can be most effectively absorbed.

Hygiene and safety are paramount in both scenarios. The injection site should be cleaned with an alcohol swab to eliminate the risk of infection, and the needle must be inserted at the correct angle—45 degrees for subcutaneous injections and 90 degrees for intramuscular injections—to ensure proper delivery of the peptide. The importance of using a new, sterile needle for each injection cannot be overstated, as this practice prevents contamination and infection, safeguarding the user's health.

After the injection, proper disposal of needles and syringes in a designated sharps container is essential for safety. This step prevents accidental needle sticks, which can transmit infections. Monitoring the injection site for signs of irritation, infection, or adverse reactions is also crucial. Any persistent redness, swelling, or discomfort should prompt consultation with a healthcare provider to address potential complications promptly.

The decision to use subcutaneous or intramuscular injections ultimately rests on the specific peptide protocol being followed and personal preference. Each method has its advantages, and understanding the nuances of both can empower users to make informed choices about their peptide therapy regimen. With careful attention to technique, hygiene, and safety, individuals can effectively administer peptides, harnessing their potential to enhance health and well-being.

3.2.3: Avoiding Injection Contamination Mistakes

Ensuring the safety and efficacy of peptide protocols involves meticulous attention to avoiding common mistakes, particularly those related to injection practices and the potential for contamination. One prevalent error is the mishandling of peptides, leading to their degradation or contamination. Peptides, being delicate in nature, require careful storage and handling. They should be stored in a cool, dry place, and once reconstituted, refrigerated and used within a recommended timeframe to maintain their potency. It's crucial to avoid exposing peptides to extreme temperatures or direct sunlight, as these conditions can break down the peptide chains, rendering them ineffective.

Another common mistake is improper reconstitution of peptides. The solvent used, typically bacteriostatic water, must be introduced slowly and gently to the peptide vial. Vigorous shaking can damage the peptide structure. Instead, gently swirl the vial to ensure the peptide dissolves properly without compromising its integrity. This method preserves the peptide's bioactivity and ensures that the dosage drawn is accurate.

Contamination is a significant risk when handling peptides, especially during the injection process. Always use sterile injection supplies and practice good hygiene. Wash hands thoroughly before handling peptides, and clean the vial's rubber stopper with an alcohol swab to kill any bacteria present. Using a new, sterile syringe and needle for each injection not only minimizes the risk of contamination but also ensures a pain-free injection as repeated use can dull the needle.

Incorrect injection technique is another area where mistakes commonly occur. Subcutaneous injections, typically administered into the fatty layer of the skin, require a small needle and precise technique to ensure the peptide is delivered to the correct tissue. Intramuscular injections, on the other hand, are delivered into the muscle and require a longer needle. Understanding the appropriate technique for each type of injection is paramount to ensure the peptide is absorbed effectively. Incorrect technique can lead to issues such as irritation at the injection site, suboptimal absorption of the peptide, and even injury.

Dosage errors can also occur, either by miscalculating the amount of peptide to reconstitute or by inaccurately measuring the dose to inject. These mistakes can be

Chapter 3:
Practical Guidelines for Peptides

mitigated by double-checking calculations and using a precision syringe that allows for accurate measurement. It's essential to start with a lower dose to assess tolerance and gradually increase as needed, rather than starting with a higher dose that could lead to adverse effects.

To avoid these common pitfalls, education and adherence to best practices in peptide handling and injection are critical. By understanding the properties of peptides, the importance of sterile technique, and the correct methods for storage, reconstitution, and administration, individuals can significantly reduce the risk of contamination and dosage errors, ensuring the safety and effectiveness of their peptide protocols.

3.3: CYCLING AND TIMING PEPTIDES FOR RESULTS

Understanding the nuances of **cycling** and **timing** your peptide regimen is crucial for harnessing their full potential. The concept of cycling involves alternating periods of peptide use with periods of abstention, allowing your body to maximize the benefits while minimizing potential desensitization to the peptides' effects. This approach mirrors the body's natural rhythms and hormonal cycles, promoting a more sustainable and effective integration of peptides into your health optimization strategies.

The **active cycle** typically ranges from 8 to 12 weeks, during which peptides are administered according to the specific protocol designed for your health goals. This is followed by a **rest period**, usually lasting 4 to 6 weeks, where peptide use is paused. This pause is vital as it helps reset your body's response to peptides, ensuring that your cells remain responsive to their regenerative and enhancing properties. It's akin to allowing the land to lie fallow in agriculture to restore its fertility.

Timing is another pivotal aspect, especially when considering the biological half-life of peptides and their peak activity periods in the body. For instance, certain peptides are more effective when taken at night due to their ability to enhance sleep quality or support the body's natural regenerative processes that occur during sleep. Others may be more beneficial when taken in the morning or before workouts, depending on their action mechanisms and your personal health objectives.

Moreover, aligning peptide administration with your body's circadian rhythms can amplify their efficacy. For example, peptides that influence growth hormone release are often best taken at night, in line with the body's natural growth hormone surge during deep sleep. Conversely, peptides that enhance energy and focus might yield the best results when taken in the morning.

The **dosage** during the active cycle should be meticulously calculated based on body weight, specific health goals, and, importantly, adjusted according to individual response and tolerance. It's essential to start with a conservative dose and gradually increase as needed, closely monitoring your body's reactions to find the optimal dosage that provides the desired effects without adverse reactions.

During the rest period, it's beneficial to focus on **nutritional support**, **exercise**, and **lifestyle adjustments** that complement the peptide therapy's goals. This holistic approach ensures that the gains made during the active cycle are maintained and even enhanced, setting a solid foundation for the next cycle.

Chapter 3:
Practical Guidelines for Peptides

Incorporating **blood work** and **biomarker tracking** can provide insights into how your body is responding to the peptides, allowing for data-driven adjustments to your protocol. This personalized approach ensures that the cycling and timing of peptide administration are finely tuned to your body's unique needs and responses, maximizing the therapeutic benefits while safeguarding against potential downsides.

By adhering to these guidelines on cycling and timing, you're not just experimenting with peptides; you're integrating them into a comprehensive, sustainable strategy for health optimization. This disciplined, informed approach ensures that you can enjoy the myriad benefits of peptides, from enhanced muscle recovery and fat loss to improved sleep and cognitive function, in the most effective and safe manner possible.

Part 2:
Targeted Health Benefits with Peptides

Chapter 4:
Peptides for Fat Loss and Metabolism

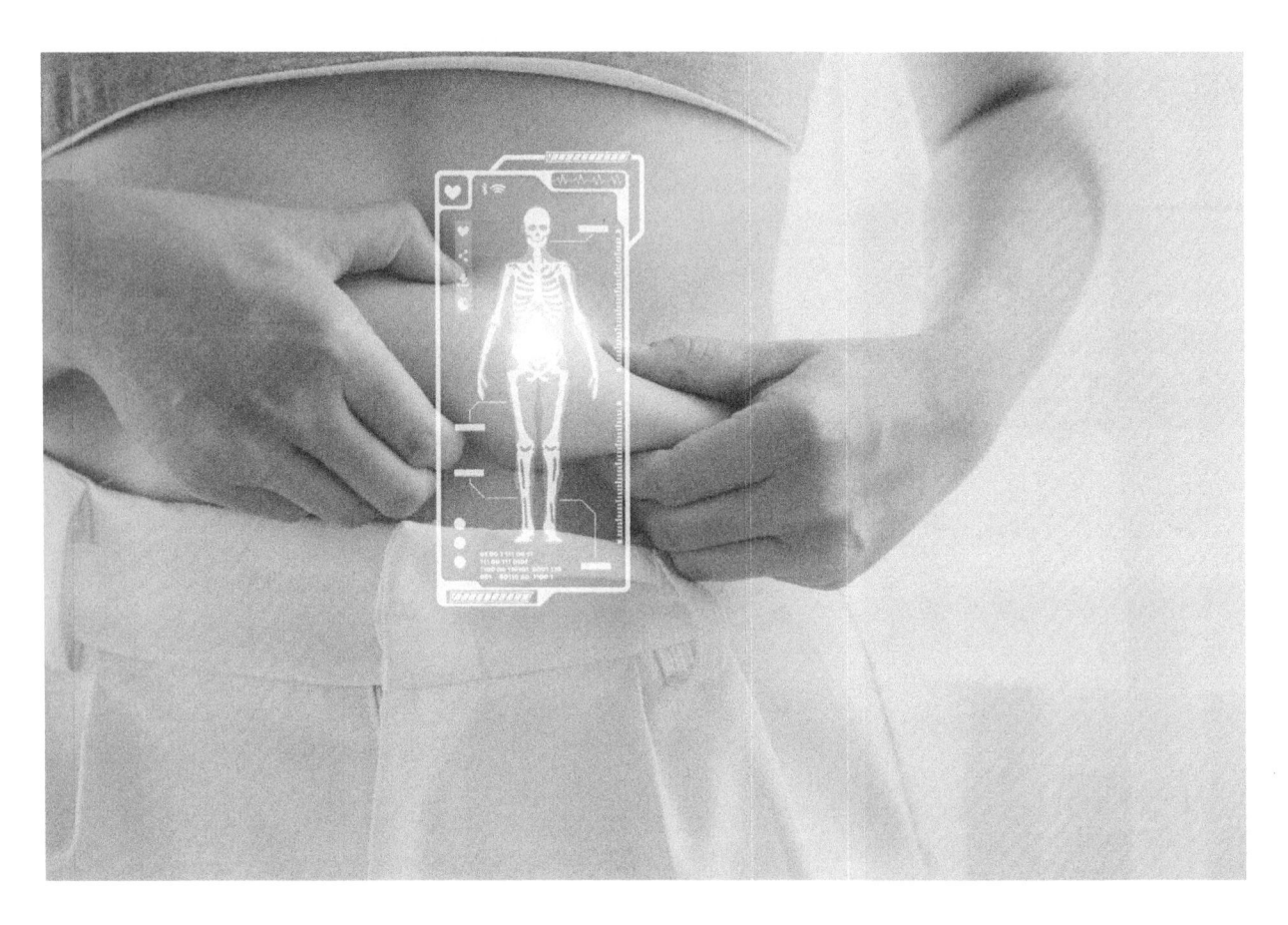

Peptides, small chains of amino acids, have emerged as powerful tools in the quest for fat loss and enhanced metabolism, offering a nuanced approach to weight management that transcends traditional diet and exercise paradigms. Their ability to modulate various physiological processes, including hormone release, tissue repair, and cellular signaling, positions them as a pivotal component in the biohacker's arsenal for achieving optimal body composition. The mechanisms through which peptides facilitate fat loss and boost metabolism are multifaceted, involving the direct stimulation of lipolysis (the breakdown of fat cells) and the enhancement of growth hormone (GH) secretion, which indirectly influences fat metabolism and muscle growth, thereby supporting a leaner physique.

AOD-9604, a peptide fragment of the C-terminal end of human growth hormone (hGH), exemplifies the direct approach to fat loss. It retains the fat-burning properties of its parent molecule without its muscle-building effects, making it an ideal candidate for those primarily interested in fat reduction. AOD-9604 operates through the stimulation of the beta-3 adrenergic receptors in adipose tissue, promoting the release of fat from fat cells and inhibiting the accumulation of new fat. This peptide has also been shown to increase the body's metabolic rate, further enhancing its fat-burning capabilities.

On the other hand, peptides like CJC-1295 and Ipamorelin represent a more indirect route to fat loss, primarily through their influence on growth hormone release. CJC-1295 is a growth hormone-releasing hormone (GHRH) analog that works by stimulating the pituitary gland to release GH, while Ipamorelin, a growth hormone-releasing peptide (GHRP), mimics the action of ghrelin, a hunger hormone that also plays a role in GH secretion. When used in combination, these peptides amplify the body's natural GH pulses, leading to increased fat metabolism, improved muscle mass, and a reduction in body fat. The synergy between CJC-1295 and Ipamorelin not only facilitates fat loss but also enhances cellular repair and regeneration, contributing to a more youthful and vigorous physique.

The strategic use of these peptides, tailored to individual needs and goals, can significantly impact one's ability to shed unwanted fat and revitalize metabolism. However, the effectiveness of peptide protocols for fat loss extends beyond the mere selection of peptides. The timing of administration, dosage, and cycle duration are critical factors that require careful consideration to maximize results while minimizing potential side effects. For instance, peptides that influence GH levels are often administered at night to coincide with the body's natural GH peak, thereby enhancing their efficacy. Dosage recommendations typically take into account factors such as body weight, desired outcomes, and previous experience with peptide therapies, underscoring the importance of a personalized approach to peptide use.

Moreover, the integration of peptide protocols into a broader lifestyle strategy encompassing diet, exercise, and sleep optimization is essential for achieving sustainable fat loss and metabolic enhancement. Peptides are not standalone solutions but rather powerful adjuncts that, when combined with healthy lifestyle choices, can significantly accelerate progress towards one's health and fitness goals.

To optimize the benefits of peptide protocols for fat loss and metabolism, adherence to a well-structured nutritional plan is paramount. A diet rich in lean proteins, healthy fats,

Chapter 4:
Peptides for Fat Loss and Metabolism

and low-glycemic carbohydrates works synergistically with peptides to fuel the body's metabolic processes and support fat loss. Incorporating intermittent fasting or time-restricted eating can further enhance the effects of peptides on metabolism, as these eating patterns have been shown to improve insulin sensitivity and increase lipolysis. The strategic timing of meals and macronutrient intake around peptide administration can also play a crucial role in maximizing their metabolic benefits. For example, consuming a protein-rich meal post-peptide injection can leverage the anabolic window for muscle repair and growth, which in turn, boosts metabolic rate.

In addition to dietary considerations, a comprehensive exercise regimen that includes both resistance training and cardiovascular activities is essential for amplifying the fat-burning and muscle-building effects of peptides. Resistance training, in particular, increases muscle mass, which naturally enhances basal metabolic rate, allowing the body to burn more calories at rest. Cardiovascular exercise, especially high-intensity interval training (HIIT), complements this by increasing calorie burn during and after workouts, further supporting fat loss efforts. The combination of targeted peptide protocols with specific exercise modalities can significantly elevate the body's fat oxidation rates, leading to more effective and sustainable body composition improvements.

Sleep optimization is another critical factor in maximizing the efficacy of peptides for fat loss and metabolism. Quality sleep supports the regulation of key hormones, including growth hormone and cortisol, which play significant roles in metabolism and fat storage. Ensuring adequate sleep duration and quality can enhance the body's responsiveness to peptides, as the majority of GH secretion occurs during deep sleep cycles. Practices such as maintaining a consistent sleep schedule, reducing blue light exposure before bedtime, and creating a restful sleeping environment can improve sleep quality, thereby supporting the hormonal balance necessary for optimal metabolic function.

Monitoring and adjusting peptide protocols based on individual response is crucial for sustained success. Regular follow-ups with a healthcare professional to assess progress and make necessary adjustments to the peptide regimen ensure that the protocols remain effective and aligned with changing health and fitness goals. This personalized approach allows for the fine-tuning of peptide types, dosages, and administration schedules to match the unique metabolic and physiological needs of each individual, optimizing the potential for fat loss and metabolic enhancement.

Finally, it's important to recognize that while peptides offer significant benefits for fat loss and metabolic health, they are most effective when used as part of a holistic approach to wellness. Combining peptide protocols with a balanced diet, regular exercise, quality sleep, and stress management practices creates a comprehensive lifestyle strategy that promotes optimal health and body composition. By focusing on these foundational aspects of health, individuals can harness the full potential of peptides to achieve their fat loss and metabolic goals, ensuring long-term success and well-being.

4.1: PEPTIDES FOR FAT BURNING AND METABOLISM

Understanding the biochemical mechanisms through which peptides promote fat burning and enhance metabolism requires a dive into the complex interplay between hormones, receptors, and cellular processes. Peptides, by their nature, are adept at targeting specific receptors on cell surfaces, initiating a cascade of events that lead to various physiological outcomes, including the mobilization of fat stores and the optimization of metabolic rates. The efficacy of peptides in fat loss and metabolic enhancement is primarily attributed to their ability to mimic or influence the action of natural hormones that regulate metabolism and energy expenditure.

One of the key hormones in this context is growth hormone (GH), which plays a pivotal role in the regulation of body composition, including the reduction of adipose tissue and the increase in lean muscle mass. Peptides such as CJC-1295 and Ipamorelin, which stimulate GH release, contribute to fat loss not only by enhancing lipolysis but also by improving metabolic functions that are conducive to weight management. The increase in GH levels, facilitated by these peptides, accelerates the breakdown of triglycerides into glycerol and free fatty acids, making these fats available for energy production and thus contributing to a reduction in body fat percentage.

Furthermore, the action of peptides on the beta-3 adrenergic receptors in adipose tissue, as seen with AOD-9604, directly stimulates fat release from fat cells. This mechanism is particularly significant because it targets fat stores for energy use without the adverse effects on insulin sensitivity or the cardiovascular system that are often associated with traditional stimulants. By activating these specific receptors, AOD-9604 promotes the utilization of fat as an energy source, enhancing metabolic rate and supporting weight loss efforts.

Another aspect of peptide-induced fat loss is the improvement of insulin sensitivity. Peptides that influence GH release also have a beneficial effect on insulin regulation, which is crucial for controlling blood sugar levels and preventing the storage of excess glucose as fat. Improved insulin sensitivity means that the body can more efficiently use glucose for energy, reducing the likelihood of glucose being converted into fat. This not

Chapter 4:
Peptides for Fat Loss and Metabolism

only aids in fat loss but also supports overall metabolic health, reducing the risk of diabetes and other metabolic disorders.

The modulation of appetite and satiety signals is another pathway through which peptides contribute to fat loss. Certain peptides can influence ghrelin levels, the hormone responsible for hunger signals, thereby reducing appetite and calorie intake. This effect, combined with the increased energy expenditure from enhanced metabolic rates, creates a favorable energy balance for fat loss. The ability of peptides to impact hunger and satiety underscores their potential as a tool for weight management, offering an advantage over traditional weight loss methods that may not address the hormonal and biochemical drivers of appetite.

Incorporating peptides into a fat loss strategy also necessitates a consideration of timing and dosage, as these factors significantly influence the peptides' effectiveness. For instance, administering peptides that stimulate GH release before sleep can align with the body's natural GH peak, enhancing the overall impact on fat metabolism and muscle repair during rest. Similarly, the dosage must be carefully calibrated to the individual's body weight, health status, and specific goals to ensure optimal results while minimizing the risk of side effects.

The integration of peptide protocols into a comprehensive approach that includes diet modification, exercise, and lifestyle adjustments further amplifies the fat-burning and metabolic benefits of peptides. A diet low in processed foods and high in protein, healthy fats, and fiber supports the actions of peptides by providing the necessary nutrients for muscle synthesis and energy production while minimizing the intake of excess calories that can lead to fat accumulation. Regular physical activity, particularly strength training and high-intensity interval training (HIIT), complements the effects of peptides on muscle growth and fat metabolism, leading to more pronounced improvements in body composition.

Ultimately, the success of peptide protocols for fat loss and metabolism enhancement hinges on a personalized approach that considers the unique physiological and metabolic characteristics of the individual. By tailoring peptide selection, dosage, and timing to the individual's needs and combining these protocols with targeted nutritional and lifestyle interventions, it is possible to maximize the fat-burning and metabolic benefits of peptides. This holistic strategy not only facilitates significant improvements in body composition but also promotes long-term health and well-being, underscoring the potential of peptides as a powerful tool in the quest for optimal physical fitness and metabolic health.

4.2: KEY PEPTIDES FOR FAT LOSS

Beyond AOD-9604 and the CJC-1295/Ipamorelin combination, several other peptides have shown promise in the realm of **fat loss** and **weight management**. **Tesamorelin**, a synthetic form of growth-hormone-releasing hormone (GHRH), has been clinically proven to reduce visceral fat in HIV patients with lipodystrophy. However, its benefits extend beyond this specific population, offering potential advantages for individuals seeking to decrease abdominal fat. Tesamorelin works by stimulating the pituitary gland to release growth hormone, which in turn, promotes lipolysis and has a positive effect on body composition. The recommended protocol involves a daily subcutaneous injection, typically in the evening, to align with the body's natural growth hormone surge. Monitoring by a healthcare professional is advisable to optimize dosage and minimize potential side effects.

Hexarelin, another potent growth hormone-releasing peptide (GHRP), exhibits a strong capacity to stimulate GH production. Its action not only facilitates fat loss but also supports muscle strength and recovery, making it a dual-purpose peptide for those looking to enhance their physique and athletic performance. Hexarelin's use, however, should be approached with caution due to its potential to increase cortisol and prolactin levels, underscoring the importance of balance and professional guidance in peptide therapy.

MT-2 (Melanotan II), primarily known for its effects on melanogenesis and sexual arousal, has also been noted to have fat-reducing properties. While not its primary use, MT-2's ability to influence appetite suppression through central melanocortin activation presents an intriguing adjunct to a comprehensive fat loss strategy. Users should be aware of its range of effects and consider them in the context of their overall health and wellness goals.

In the pursuit of fat loss, **fragment 176-191**, a fragment of the HGH molecule, has garnered attention for its ability to specifically target adipose tissue without the broader effects on insulin sensitivity or IGF-1 levels that full-length HGH might induce. This specificity makes fragment 176-191 an attractive option for those focused solely on fat loss, with a protocol that often involves twice-daily injections to maintain steady plasma levels.

The integration of these peptides into a fat loss regimen should be done with a holistic perspective, considering not only the direct effects of the peptides themselves but also how they fit into a broader lifestyle approach that includes diet, exercise, and sleep optimization. For example, coupling peptide use with a high-protein, low-glycemic diet can enhance the peptides' fat-burning effects, while resistance training and cardio can amplify growth hormone's impact on fat loss and muscle growth. Furthermore, optimizing sleep quality can improve the body's natural growth hormone production, creating a conducive environment for peptides to exert their maximum benefit.

It's also critical to approach peptide therapy with a mindset of patience and consistency. Results from peptide use, particularly in the context of fat loss, are not instantaneous but rather cumulative, requiring adherence to a protocol over weeks or months. Regular consultation with a healthcare provider can help tailor the peptide regimen to the individual's changing needs and circumstances, ensuring that the strategy remains effective and aligned with health goals.

In conclusion, the landscape of peptides for fat loss is rich and varied, offering multiple avenues for individuals to explore in their quest for improved body composition. Whether considering AOD-9604, CJC-1295/Ipamorelin, Tesamorelin, Hexarelin, MT-2, or fragment 176-191, the key to success lies in a personalized approach that integrates these powerful tools into a comprehensive lifestyle strategy. With the right combination of peptides, diet, exercise, and lifestyle adjustments, achieving and maintaining optimal fat loss is an attainable goal.

4.2.1: AOD-9604 Fat Reduction Protocols

AOD-9604 stands out in the realm of fat reduction peptides due to its unique mechanism of action, which targets fat cells directly without the broad systemic effects associated with human growth hormone (hGH). This peptide fragment has been meticulously engineered to harness the fat-burning prowess of hGH's C-terminal end, focusing specifically on adipose tissue reduction. AOD-9604 activates the beta-3 adrenergic receptors in fat cells, a critical pathway that signals the breakdown of fat (lipolysis) and inhibits the formation of new fat cells (lipogenesis). This dual action not only accelerates fat loss but also prevents future fat accumulation, offering a sustainable approach to weight management.

The metabolic enhancement attributed to AOD-9604 extends beyond simple fat reduction. By stimulating lipolysis without significantly impacting blood sugar levels or increasing insulin resistance, AOD-9604 offers a weight loss solution that circumvents the common pitfalls associated with other fat-burning interventions. Its ability to increase the body's metabolic rate further amplifies its fat-reducing effects, making it an invaluable tool for individuals aiming to improve their body composition.

For those integrating AOD-9604 into their health and fitness regimen, the protocol typically involves subcutaneous injections, as this method ensures optimal bioavailability and efficacy. The recommended dosage ranges from 120 to 300 micrograms per day, divided into one or two doses, depending on individual goals and response. Starting at the lower end of the dosage spectrum and gradually increasing allows for personalized optimization, minimizing the risk of adverse effects while maximizing fat loss benefits.

Timing of administration plays a pivotal role in AOD-9604's effectiveness. For enhanced results, injections should ideally be administered in the morning or before physical activity. Morning administration leverages the body's natural circadian rhythms, optimizing metabolic rate throughout the day. Pre-activity injections, on the other hand, capitalize on the increased metabolic demand of exercise, further boosting fat oxidation rates.

It is crucial for users to monitor their response to AOD-9604 closely, adjusting dosages as necessary to align with their evolving fitness goals and physiological responses. Regular consultations with a healthcare provider knowledgeable in peptide therapies can provide tailored guidance, ensuring safe and effective use. As with any peptide protocol, the purity and source of AOD-9604 are paramount. Only purchase from reputable suppliers who provide third-party testing results to verify purity and potency, safeguarding against contaminants and substandard products that could undermine health and progress.

Incorporating AOD-9604 into a broader lifestyle strategy that emphasizes balanced nutrition, regular exercise, and adequate rest enhances its fat reduction capabilities. A nutrient-dense diet, rich in proteins, healthy fats, and complex carbohydrates, supports metabolic health and complements the fat-burning effects of AOD-9604. Coupling peptide therapy with strength training and cardiovascular exercise maximizes muscle preservation and fat loss, leveraging the synergistic effects of AOD-9604 on metabolism and body composition. Furthermore, prioritizing sleep and stress management optimizes hormonal balance, creating an internal environment conducive to weight loss and overall well-being.

The strategic use of AOD-9604, when aligned with a comprehensive approach to health and fitness, empowers individuals to achieve and maintain their desired body composition. By understanding the specific actions and benefits of AOD-9604, adhering to recommended protocols, and integrating supportive lifestyle practices, users can unlock the full potential of this peptide in their journey towards optimal health and body aesthetics.

Chapter 4:
Peptides for Fat Loss and Metabolism

4.2.2: CJC-1295 and Ipamorelin Synergy

The combination of CJC-1295 and Ipamorelin represents a powerful synergistic approach to fat loss, leveraging the unique properties of each peptide to amplify results. CJC-1295, a long-acting Growth Hormone Releasing Hormone (GHRH) analog, works to increase plasma growth hormone levels, while Ipamorelin, a selective Growth Hormone Secretagogue (GHS), acts on the ghrelin receptor to stimulate the release of growth hormone by the pituitary gland. This combination not only enhances growth hormone levels in the body but does so in a way that mimics natural physiological processes, leading to more sustained and consistent growth hormone pulses that are crucial for fat metabolism and overall weight management.

For individuals seeking to utilize this combination for fat loss, it is essential to understand the dosing protocols that have been found to be most effective. Typically, CJC-1295 is administered once weekly due to its long half-life, ensuring a steady release of growth hormone. The recommended dose ranges from 1 to 2 mg per injection, depending on individual factors such as body weight, metabolic rate, and specific fat loss goals. On the other hand, Ipamorelin's shorter half-life necessitates more frequent administration, often recommended as 200 to 300 mcg three times daily. This regimen ensures that growth hormone levels remain elevated throughout the day, optimizing fat metabolism and supporting weight loss efforts.

The cycle duration for a CJC-1295 and Ipamorelin combination typically spans between 8 to 12 weeks, followed by a 4 to 6-week break. This cycling strategy not only prevents the possible desensitization of pituitary cells but also aligns with the body's natural rhythms, allowing for periods of enhanced growth hormone release followed by necessary rest and recovery phases. During the active cycle, it's crucial to monitor the body's response to the therapy, adjusting dosages as needed to optimize results and minimize potential side effects.

To maximize the fat loss benefits of the CJC-1295 and Ipamorelin combination, integrating these peptides with a comprehensive lifestyle approach is advisable. This includes maintaining a balanced diet rich in nutrients that support metabolic health, engaging in regular physical activity that includes both strength training and cardiovascular exercises, and ensuring adequate sleep to facilitate recovery and hormonal balance. Additionally, staying hydrated and managing stress levels can further enhance the efficacy of peptide therapy for fat loss.

It's also important for individuals to consult with healthcare professionals experienced in peptide therapy to tailor the protocol to their specific needs and circumstances. This personalized approach ensures not only the safety and effectiveness of the treatment but also aligns with the individual's health status, goals, and lifestyle factors. Regular monitoring through blood work and other relevant biomarkers is recommended to track progress and make necessary adjustments to the protocol for optimal results.

In conclusion, the strategic use of CJC-1295 and Ipamorelin in combination offers a promising approach to enhancing fat loss efforts. By understanding the nuances of dosing, cycling, and integrating these peptides with a healthy lifestyle, individuals can harness the synergistic effects of these compounds to achieve their weight management goals.

4.3: PROTOCOLS FOR EFFECTIVE WEIGHT MANAGEMENT

Beyond the initial combination of CJC-1295 and Ipamorelin, effective weight management protocols can be further optimized by incorporating additional peptides known for their fat loss properties. **Tesamorelin**, for instance, has shown significant efficacy in reducing visceral fat, making it a valuable addition to a comprehensive weight management strategy. When combined with the CJC-1295 and Ipamorelin duo, Tesamorelin can be administered at a dose of 2mg daily for a period of 3 months, followed by a reassessment of body composition to adjust dosages as necessary. This approach allows for a targeted attack on stubborn fat areas, enhancing the overall fat loss effect.

Dietary adjustments play a critical role in supporting the fat loss induced by peptide protocols. A diet low in processed sugars and high in protein, healthy fats, and fiber aids in optimizing the hormonal environment for fat loss. It's crucial to align meal timing with peptide administration to leverage the peaks in growth hormone release for maximum fat metabolism. For instance, consuming a protein-rich meal post-peptide injection can amplify the body's natural growth hormone response, further supporting muscle synthesis and fat loss.

Physical activity is another cornerstone of effective weight management with peptides. A combination of resistance training and high-intensity interval training (HIIT) complements the fat loss and muscle-building effects of peptides. Resistance training should be prioritized on days of peptide administration, particularly after injections of CJC-1295 and Ipamorelin, to take advantage of the increased growth hormone levels. On rest days or during the off-cycle periods, maintaining a consistent level of moderate activity helps sustain metabolic rate and supports continued fat loss.

Chapter 4:
Peptides for Fat Loss and Metabolism

Hydration and sleep are often overlooked aspects of weight management but are vital when using peptides. Adequate water intake ensures optimal cellular function and aids in the efficient metabolism of fat. Similarly, quality sleep is essential for the body to repair and build muscle tissue, as well as to maintain a healthy balance of hunger and satiety hormones. Ensuring 7-9 hours of uninterrupted sleep aligns with the body's natural growth hormone release cycles, maximizing the regenerative and fat-burning processes that occur during deep sleep stages.

Monitoring and adjustment of the protocol are essential for sustained success in weight management. Regular check-ins with a healthcare professional experienced in peptide therapy allow for the fine-tuning of dosages and the introduction of new peptides or supplements based on the individual's progress and any side effects experienced. Blood work and body composition analysis provide objective measures of the protocol's effectiveness and guide adjustments to dosages, cycling patterns, and lifestyle factors.

Incorporating these strategies into a cohesive protocol maximizes the potential for fat loss and muscle preservation, leading to improved body composition and overall health.

Chapter 5: Muscle Building with Peptides

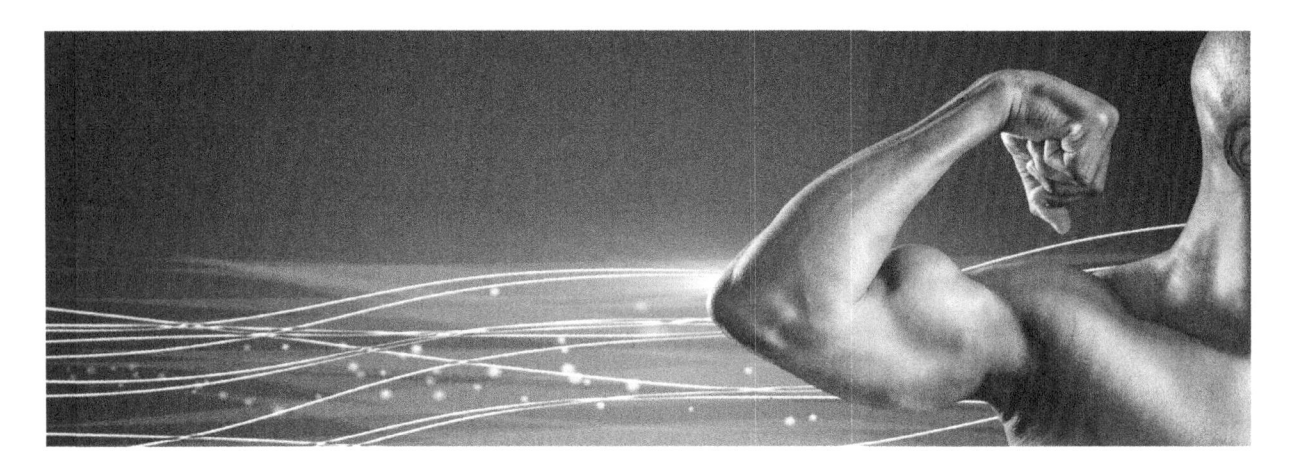

Peptides, short chains of amino acids, have emerged as a revolutionary tool in enhancing muscle growth and expediting recovery times for athletes and fitness enthusiasts alike. Their ability to stimulate the human growth hormone (HGH) secretion plays a pivotal role in muscle synthesis and repair, making them an invaluable component of any serious training regimen. GHRP-2 and GHRP-6, for instance, are two peptides known for their potent effects on growth hormone release. They operate by mimicking the action of ghrelin, a hunger-stimulating hormone that also has a significant impact on the body's growth hormone levels. When administered, these peptides can increase appetite as a side effect, but more importantly, they amplify the natural production of growth hormone, thereby facilitating more effective muscle growth and faster recovery from intense physical activity.

The role of peptides in muscle recovery cannot be overstated. After strenuous workouts, the body needs to repair itself and adapt to the stresses it has been subjected to. Peptides like IGF-1 LR3 play a crucial role in this process. IGF-1, or Insulin-like Growth Factor 1, closely mimics the insulin hormone and has profound effects on muscle regeneration. It promotes nitrogen retention and protein synthesis, which are essential for muscle repair and growth. This peptide not only helps in building new muscle fibers but also aids in the recovery of damaged ones, making it an essential tool for athletes looking to enhance their performance and reduce downtime due to injuries.

Moreover, the strategic use of peptide protocols can significantly improve the efficacy of muscle building and recovery processes. By understanding the specific actions and benefits of different peptides, individuals can tailor their peptide regimen to align with their fitness goals and physical condition. For instance, combining GHRP-6 with CJC-1295, a GHRH (Growth Hormone-Releasing Hormone) analog, can synergistically enhance growth hormone levels beyond what each peptide could achieve alone. This combination not only maximizes the growth hormone pulse but also extends its duration, providing a conducive environment for muscle growth and recovery over a more extended period.

It is also essential to consider the timing of peptide administration in relation to exercise routines. Administering peptides like GHRP-2 or IGF-1 LR3 immediately before or after workouts can optimize the body's response to the peptides, taking advantage of the natural increase in growth hormone levels that exercise induces. This timing ensures that the body has the necessary resources to begin the repair and growth process as soon as the workout is completed, thereby reducing recovery times and enhancing overall muscle growth.

The benefits of peptides extend beyond mere muscle synthesis and recovery. They also play a crucial role in reducing inflammation, improving joint health, and enhancing the body's ability to recover from injuries. Peptides like BPC-157, known for its regenerative properties, can significantly improve recovery times from muscle sprains, tears, and other injuries, further underscoring the versatility and efficacy of peptides in a comprehensive fitness and wellness strategy.

Incorporating peptides into a muscle-building and recovery protocol requires a nuanced understanding of each peptide's unique properties and how they interact with the body's physiological processes. It is crucial to start with lower dosages, gradually increasing them to gauge the body's response and minimize potential side effects. Consulting with healthcare professionals experienced in peptide therapy can provide personalized guidance and ensure that the chosen peptide regimen is safe, effective, and aligned with the individual's health status and fitness goals.

As athletes and fitness enthusiasts aim to push their limits, the role of peptides in enhancing endurance and reducing fatigue becomes increasingly important. Peptides such as **TB-500**, known for its ability to promote cell growth, wound healing, and increase endurance, offer a significant advantage in extending the capacity for physical exertion and improving overall athletic performance. This peptide facilitates the development of new blood vessels, a process known as angiogenesis, and the production of new blood cells, which are crucial for delivering oxygen and nutrients to muscles, thereby enhancing stamina and endurance during prolonged periods of physical activity.

Furthermore, peptides like **MK-677**, which acts as a potent, orally active growth hormone secretagogue, mimics the GH stimulating action of the endogenous hormone ghrelin. It has been shown to increase lean body mass and decrease fat mass. MK-677 is particularly beneficial for those looking to improve their physique by enhancing muscle mass and reducing body fat, all while promoting deeper, more restorative sleep. Sleep is a critical component of the recovery process, as it is during this time that the body undergoes most of its repair and regeneration. By improving sleep quality, MK-677 indirectly supports muscle recovery and growth, making it a dual-purpose peptide for those focused on both performance and recovery.

The integration of peptides into a fitness regimen must also take into account the importance of hydration, nutrition, and rest. Adequate hydration is essential for optimal peptide function, as it supports cellular health and ensures the efficient transport of peptides to their target sites. A balanced diet rich in proteins, healthy fats, and carbohydrates provides the necessary building blocks for muscle repair and growth, while also supporting the body's response to peptide therapy. Additionally, allowing sufficient time for rest and recovery is crucial in preventing overtraining and ensuring that the body has the opportunity to fully benefit from the regenerative properties of peptides.

To further enhance the benefits of peptide therapy, individuals may consider incorporating **antioxidant-rich foods and supplements** into their diet to combat oxidative stress and inflammation, which can impede recovery and muscle growth. Antioxidants such as vitamin C, vitamin E, and glutathione not only support the body's natural defense mechanisms but also improve the efficacy of peptides in promoting muscle repair and growth.

In conclusion, the strategic application of peptide therapy offers a promising avenue for those seeking to optimize muscle building and recovery. By carefully selecting peptides based on their specific benefits, considering the timing of administration, and integrating these compounds into a holistic approach that includes proper nutrition, hydration, and rest, individuals can maximize their fitness outcomes. The potential of peptides to enhance muscle synthesis, accelerate recovery, improve endurance, and support overall health underscores their value in a comprehensive fitness and wellness strategy. With a commitment to understanding and applying these principles, athletes and fitness enthusiasts can unlock the full potential of peptide therapy to achieve their performance and physique goals.

5.1: MUSCLE SYNTHESIS AND RECOVERY MECHANISMS

The intricate process of muscle synthesis and recovery is fundamentally driven by the delicate balance of protein synthesis and breakdown within the muscle cells. Peptides, by their very nature, play a pivotal role in tipping this balance towards synthesis, facilitating muscle growth and expediting recovery. The cellular mechanisms underlying these processes are complex, involving a multitude of signaling pathways and molecular interactions that are precisely orchestrated to achieve the desired outcomes of enhanced muscle mass and rapid recovery from exercise-induced damage.

At the heart of muscle synthesis is the mTOR pathway, a critical regulator of cellular growth and protein synthesis. Peptides such as IGF-1 and mechano-growth factor (MGF) directly stimulate the mTOR pathway, thereby enhancing protein synthesis. IGF-1, for instance, binds to its receptor on muscle cells, triggering a cascade of intracellular signaling that culminates in the activation of mTOR. This activation not only promotes the assembly of protein-building machinery but also increases the uptake of amino acids into the cells, providing the necessary building blocks for muscle growth. MGF, on the other hand, is expressed in response to mechanical stress on muscles, such as that experienced during resistance training. It acts locally at the site of muscle damage, initiating the repair process by stimulating satellite cell activation and proliferation. These satellite cells are crucial for muscle regeneration, as they fuse with damaged muscle fibers, donating their nuclei to support the repair and synthesis of new muscle proteins.

Recovery from exercise-induced muscle damage is equally critical to the muscle-building process. Peptides such as BPC-157 and TB-500 offer significant benefits in this regard, primarily through their anti-inflammatory and regenerative properties. BPC-157, a peptide with a strong track record of accelerating wound healing, has been shown to significantly reduce inflammation in damaged tissues, thereby alleviating pain and promoting faster recovery. Its mechanisms of action include the upregulation of growth hormone receptors and the modulation of nitric oxide synthesis, which together enhance blood flow to the injured area, supplying the nutrients and oxygen required for tissue repair. TB-500, known for its ability to increase cell migration and proliferation, plays a vital role in the regeneration of injured muscle fibers. By stimulating the production of actin, a protein that forms the contractile filaments of muscle cells, TB-500 facilitates the remodeling of the cytoskeleton, which is essential for cell migration, proliferation, and ultimately, the healing of muscle tissue.

The timing of peptide administration is critical to maximizing their muscle-building and recovery-enhancing effects. Administering growth hormone-releasing peptides such as GHRP-2 or GHRP-6 before workouts can amplify the body's natural growth hormone surge induced by exercise, providing a more robust environment for muscle growth. Post-workout, peptides like IGF-1 or MGF can be particularly effective, as they capitalize on the increased blood flow to muscles during this period, ensuring efficient delivery to the sites where they are most needed.

In addition to direct interventions, supporting the body's natural recovery processes is essential for optimal muscle synthesis and repair. Adequate hydration, nutrition, and sleep are foundational elements that complement peptide therapy. A diet rich in proteins and amino acids provides the raw materials for muscle repair, while hydration ensures that nutrients are efficiently transported throughout the body. Sleep, particularly deep sleep, is a critical period for muscle recovery, as growth hormone levels peak during this time, facilitating the repair and growth of muscle tissue.

Understanding the cellular mechanisms behind muscle synthesis and recovery allows for a more targeted approach to peptide therapy, enabling individuals to select peptides that align with their specific goals and needs. By leveraging the unique properties of different peptides and optimizing their administration in conjunction with lifestyle factors, individuals can significantly enhance their muscle growth and recovery, pushing the boundaries of their physical performance and achieving their fitness objectives.

5.2: TOP PEPTIDES FOR MUSCLE GROWTH

TB-500, also known as Thymosin Beta-4, is widely recognized for its profound healing properties, but its role in muscle growth and performance enhancement is equally significant. TB-500 facilitates the process of angiogenesis, the formation of new blood vessels, and upregulates the cell-building protein, actin. These functions are crucial for athletes as they not only improve wound healing and reduce inflammation but also significantly contribute to muscle growth and endurance. The recommended dosage for TB-500 ranges from 2 to 2.5 mg taken twice weekly over a period of 4 to 6 weeks. This regimen can lead to noticeable improvements in flexibility, reduced inflammation, and enhanced muscle growth. It's important for users to start with the lower end of the dosage range to assess tolerance and gradually increase as needed.

MK-677, or Ibutamoren, stands out for its unique ability to mimic the action of ghrelin and increase growth hormone levels without altering cortisol levels. This property makes MK-677 an exceptional choice for those looking to enhance muscle mass, improve sleep quality, and accelerate recovery times. For muscle growth and performance enhancement, a daily dose of 25 mg of MK-677 before bed can significantly increase growth hormone levels, promoting anabolic growth and fat loss. Users often report improved muscle density, enhanced recovery, and increased appetite, which can be beneficial for those in a bulking phase.

BPC-157 is another peptide that has gained attention for its remarkable healing properties, particularly in the context of tendon and ligament repair. However, its benefits extend to muscle growth and recovery, making it a versatile peptide for athletes. BPC-157 enhances the body's natural healing processes, leading to faster recovery from muscle tears and injuries. This peptide can be administered at a dose of 200 to 800 mcg per day, depending on the severity of the injury and the desired outcome. Athletes using BPC-157 often experience rapid recovery times, allowing for more intense and frequent training sessions without the typical risk of overuse injuries.

Hexarelin, a potent GHRP (Growth Hormone Releasing Peptide), stimulates the release of growth hormone, leading to increased muscle mass, strength, and recovery. Unlike other GHRPs, Hexarelin does not significantly increase appetite, making it an ideal choice for athletes focused on lean muscle gains. A typical dosage of Hexarelin is 100 mcg administered three times daily. This peptide is known for its rapid onset of action, with users reporting significant improvements in strength and muscle fullness within weeks of starting the protocol.

Incorporating these peptides into a comprehensive training and nutrition plan can significantly enhance muscle growth, recovery, and athletic performance. It's crucial for individuals to monitor their body's response to these peptides closely and adjust dosages as necessary. Consulting with a healthcare professional experienced in peptide therapy can provide personalized guidance and ensure the safety and effectiveness of the treatment. Additionally, combining peptide therapy with adequate nutrition, hydration, and rest will maximize the benefits and support long-term health and performance goals.

5.2.1: GHRP-2 and GHRP-6 for Muscle Growth

GHRP-2 and GHRP-6, both growth hormone-releasing peptides, stand at the forefront of facilitating muscle growth and enhancing recovery processes. Their mechanism of action primarily involves stimulating the pituitary gland to release growth hormone (GH), which plays a pivotal role in muscle development and the body's recovery from physical stress. The efficacy of these peptides in muscle growth is attributed to their ability to mimic the ghrelin hormone, which not only influences GH release but also

Chapter 5:
Muscle Building with Peptides

impacts appetite, fat metabolism, and energy homeostasis. This dual action makes GHRP-2 and GHRP-6 particularly effective for individuals aiming to increase muscle mass while simultaneously reducing body fat.

The dosing guidelines for GHRP-2 and GHRP-6 are critical to achieving optimal results. For GHRP-2, the recommended starting dose is typically around 100 mcg per injection, administered two to three times daily. This dosing frequency capitalizes on the natural pulsatile release of growth hormone, aligning with the body's circadian rhythm to maximize the anabolic effects while minimizing potential side effects. It's essential to inject on an empty stomach or with only protein in the system to enhance the peptide's effectiveness, as carbohydrates and fats can blunt the GH release stimulated by GHRP-2.

GHRP-6, while similar in function to GHRP-2, may induce a stronger increase in appetite due to its profound ghrelin mimicking effect. This characteristic can be advantageous for individuals struggling to meet their caloric intake requirements for muscle gain. The recommended dosage for GHRP-6 mirrors that of GHRP-2, with injections of 100 mcg two to three times daily. Adjustments to dosage should be made based on individual response, goals, and tolerance, with careful monitoring for any signs of side effects such as water retention or increased hunger levels.

Combining GHRP-2 or GHRP-6 with a GHRH (Growth Hormone Releasing Hormone) analogue such as CJC-1295 can synergistically amplify the GH pulse, further enhancing muscle growth and recovery. This combination approach, often referred to as a 'stack', leverages the complementary mechanisms of action of GHRPs and GHRH analogues to maximize GH release. A common protocol involves administering GHRP-2 or GHRP-6 alongside CJC-1295, with dosages adjusted to suit individual tolerance and goals. The CJC-1295 is typically dosed at 100 mcg per injection, with the timing aligned with GHRP-2 or GHRP-6 injections to exploit the synergistic GH release.

It's imperative for users to source these peptides from reputable suppliers to ensure purity and efficacy. Substandard products not only jeopardize health but also compromise the potential benefits of the therapy. Additionally, individuals should approach peptide use with a holistic perspective, incorporating adequate nutrition, exercise, and rest into their regimen to support the physiological processes influenced by GH.

The administration of GHRP-2 and GHRP-6 should be approached with precision, respecting the body's natural rhythms and hormonal balance. Users are advised to start with the lower end of the dosage range, gradually increasing as needed while closely monitoring the body's response. This cautious approach allows for the identification of the optimal dose that provides the desired anabolic benefits without eliciting adverse effects.

In the context of muscle growth and recovery, the strategic use of GHRP-2 and GHRP-6, particularly when combined with a GHRH analogue, offers a potent tool for enhancing physical performance and achieving body composition goals. However, the success of peptide therapy hinges on adherence to dosing protocols, quality of peptides, and a comprehensive lifestyle approach that fosters an optimal environment for growth and recovery.

5.2.2: IGF-1 LR3 for Muscle Enhancement

IGF-1 LR3, a potent variant of the naturally occurring IGF-1 (Insulin-like Growth Factor 1), has garnered significant attention within the biohacking community for its muscle-enhancing properties. This longer-acting version of IGF-1 is engineered to resist deactivation by IGF-1 binding proteins in the bloodstream, thereby extending its efficacy period and enhancing its muscle-building potential. The primary mechanism through which IGF-1 LR3 operates is by binding to the IGF-1 receptor present on muscle cells, initiating intracellular signaling that promotes protein synthesis and discourages protein degradation. This dual action not only accelerates muscle growth but also plays a crucial role in muscle recovery, making it an invaluable tool for athletes and individuals engaged in regular physical training.

The benefits of IGF-1 LR3 extend beyond mere muscle augmentation; it also contributes to the improvement of muscle recovery times, reduction in body fat, enhanced performance, and overall vitality. Its ability to stimulate the proliferation of cells (hyperplasia) sets it apart from other growth factors that merely increase the size of existing cells (hypertrophy). This unique feature of IGF-1 LR3 allows for the development of new muscle fibers, offering long-term benefits in muscle density and strength.

For those considering the incorporation of IGF-1 LR3 into their peptide protocol, it is imperative to understand the recommended modalities of use to maximize its benefits while minimizing potential risks. The dosage of IGF-1 LR3 typically ranges from 20 to 120 mcg per day, with a sweet spot for most individuals lying around 40 to 80 mcg. It is administered through subcutaneous injections, and due to its potent nature, a cycle should not exceed 4 to 6 weeks, followed by a rest period to allow the body to normalize its IGF-1 sensitivity. Timing the injections post-workout can leverage the body's natural insulin spike, enhancing the uptake of IGF-1 LR3 into muscle cells and amplifying its anabolic effects.

Given the powerful effects of IGF-1 LR3 on muscle growth and recovery, it is also crucial to pair its use with a well-structured training program and a nutrient-dense diet. Adequate protein intake is essential to supply the building blocks for muscle synthesis, while sufficient calories support overall growth and recovery. Hydration and rest cannot be overstated, as they are fundamental to optimizing the body's response to IGF-1 LR3 and ensuring sustainable progress.

While IGF-1 LR3 presents a promising avenue for muscle enhancement, individuals must approach its use with caution. Consulting with a healthcare professional experienced in peptide therapy is advisable to tailor the protocol to one's specific health profile and goals. Regular monitoring through blood work can help track the body's response and adjust dosages as needed, ensuring safety and efficacy throughout the cycle.

In conclusion, IGF-1 LR3 offers a powerful tool for those seeking to enhance muscle growth, recovery, and overall physical performance. By understanding its mechanisms, benefits, and recommended usage, individuals can safely integrate this peptide into their health optimization regimen, unlocking new potentials in their fitness journey.

5.3: PEPTIDE PROTOCOLS FOR ACCELERATED RECOVERY

Building on the foundation laid out for understanding IGF-1 LR3's role in muscle growth and recovery, it's crucial to explore additional peptide combinations that can further enhance recovery times and muscle repair. **Peptide stacking** is a strategic approach to combine different peptides, each with unique properties, to create a synergistic effect that can significantly accelerate the recovery process post-intense workouts. This method allows for a more comprehensive recovery strategy, addressing various aspects of muscle repair, inflammation reduction, and overall recovery speed.

One effective stack combines **BPC-157** and **TB-500**, both renowned for their healing properties. BPC-157, a Body Protecting Compound, has been shown to significantly accelerate wound healing, including muscle and tendon tissues, by promoting the formation of new blood vessels and facilitating the migration of tendon fibroblasts. This action not only enhances recovery but also reduces the risk of future injuries by strengthening the muscle fibers and connective tissues. On the other hand, TB-500, a Thymosin Beta-4 peptide, plays a crucial role in reducing inflammation, improving flexibility, and promoting the growth of new blood cells and muscle tissue. When used together, BPC-157 and TB-500 create a powerful recovery stack that not only speeds up the healing process but also improves the quality of the muscle tissue, making it more resilient to strain and injury.

For dosage, starting with **BPC-157 at 200-800 mcg per day** and **TB-500 at 2-5 mg per week** is recommended, adjusting based on individual response and specific recovery needs. It's important to note that while these dosages serve as a general guideline, consulting with a healthcare professional experienced in peptide therapy is essential to tailor the protocol to your unique health profile and recovery goals.

Another noteworthy combination for accelerated recovery involves **MK-677**, also known as Ibutamoren, which acts as a growth hormone secretagogue. MK-677 increases the secretion of growth hormone and IGF-1, both vital for muscle repair and recovery. By elevating growth hormone levels, MK-677 enhances muscle growth, improves sleep quality—a critical component of the recovery process—and boosts overall recovery speed. Combining MK-677 with the previously mentioned peptides can further optimize recovery by providing a multifaceted approach that addresses muscle repair, inflammation, and growth hormone levels.

Starting with an **MK-677 dosage of 10-25 mg per day** can provide significant benefits in recovery and muscle growth. However, due to its potential to increase appetite, monitoring weight and adjusting the dosage accordingly is advisable to ensure optimal results without unwanted side effects.

Incorporating these peptides into a well-designed recovery protocol requires careful consideration of timing, dosage, and cycling. Administering peptides post-workout or before bed can optimize their recovery benefits, taking advantage of the body's natural repair processes during sleep. Cycling these peptides, using them for a period followed by a rest period, can help maintain their efficacy and minimize the risk of desensitization.

Ultimately, the goal of peptide stacking for accelerated recovery is to create a comprehensive, personalized protocol that supports rapid healing, reduces downtime, and enhances overall performance. By carefully selecting and combining peptides based on their unique properties and recovery goals, individuals can significantly improve their recovery times, ensuring they return to training stronger and more resilient. Regular monitoring and adjustments to the protocol, guided by a healthcare professional's expertise, will ensure safety and maximize the benefits of peptide therapy for accelerated recovery.

5.3.1: Peptide Stacking for Muscle Repair

Expanding upon the foundational knowledge of peptide stacking for muscle repair, it's essential to delve deeper into the nuanced strategies that can optimize the healing and strengthening of muscle tissues. The combination of **BPC-157** and **TB-500** serves as a cornerstone for muscle repair protocols, but the integration of additional peptides can further enhance recovery and fortify muscle integrity. **GHK-Cu**, a peptide known for its tissue regeneration and anti-inflammatory properties, emerges as a pivotal addition to the muscle repair stack. When GHK-Cu is introduced into the regimen, it not only accelerates the repair of damaged muscle fibers but also stimulates the production of collagen and elastin, crucial components for maintaining the elasticity and strength of muscle tissues.

The dosing strategy for GHK-Cu typically involves a range of **2-10 mg per week**, administered via subcutaneous injections. This peptide, renowned for its ability to modulate copper in the body, plays a vital role in the body's healing process, making it an invaluable asset in the context of muscle repair. The inclusion of GHK-Cu in a peptide stack aimed at muscle repair should be carefully calibrated, taking into consideration the individual's specific recovery needs and the synergistic effects of combining it with BPC-157 and TB-500. It's advisable to start at the lower end of the dosing spectrum and gradually adjust based on the body's response and the progression of muscle repair.

Another critical component of an advanced muscle repair stack is **MGF (Mechano Growth Factor)**, a peptide that specifically targets skeletal muscle repair and growth. MGF is unique in its ability to stimulate muscle growth through the activation of muscle stem cells, promoting the regeneration of muscle tissue and providing a potent anabolic effect. The recommended dosage for MGF ranges from **200-400 mcg per session**, administered post-workout to capitalize on the body's natural growth signals. The timing of MGF injections is crucial, as its effects are maximally potent when muscle tissues are primed for repair following exercise-induced stress.

For individuals seeking to maximize muscle repair and recovery, the orchestration of peptide timing, dosage, and combination becomes a finely tuned symphony of biological processes. The post-workout window is particularly critical for peptide administration, as this is when the body's natural repair mechanisms are most active. Leveraging this window with precise peptide protocols can significantly enhance the effectiveness of muscle repair and growth strategies.

Moreover, the concept of **peptide cycling** plays a pivotal role in optimizing the body's response to these powerful molecules. Implementing cycles of 4-6 weeks of intensive peptide therapy followed by periods of rest allows the body to maximize the benefits of peptides without becoming desensitized to their effects. This cycling strategy ensures that muscle repair and growth are not only accelerated but also sustained over time, leading to lasting improvements in muscle strength and resilience.

Incorporating peptides such as BPC-157, TB-500, GHK-Cu, and MGF into a comprehensive muscle repair protocol requires a holistic approach that considers not only the peptides themselves but also the individual's overall health, activity level, and recovery goals. Tailoring the combination and dosage of peptides to the individual's unique physiology and recovery needs is essential for achieving optimal results. Regular consultation with healthcare professionals experienced in peptide therapy, along with ongoing monitoring of progress and adjustments to the protocol, ensures that the muscle repair strategy is both effective and safe.

As the science of peptide therapy continues to evolve, so too do the strategies for harnessing these potent molecules for muscle repair and recovery. By staying informed about the latest advancements in peptide research and adhering to a disciplined approach to peptide stacking, individuals can significantly enhance their ability to recover from injuries, improve muscle performance, and achieve their health and fitness goals.

5.3.2: Enhancing Recovery for High-Intensity Training

For individuals engaging in high-intensity training, the recovery period is as crucial as the training itself. Accelerating recovery times ensures that you can return to your training sessions sooner, with reduced risk of injury and improved performance. A strategic approach to peptide use can significantly enhance this recovery process, particularly when tailored to the demands of high-intensity workouts. Peptides such as **BPC-157** and **TB-500**, known for their healing properties, are foundational. However, integrating additional peptides like **MGF (Mechano Growth Factor)** and **GHK-Cu** can provide a more comprehensive recovery solution, addressing not only muscle repair but also inflammation and overall tissue health.

MGF should be administered post-workout to exploit the body's natural repair mechanisms activated by exercise-induced stress. This timing ensures that MGF's muscle stem cell-stimulating properties are fully leveraged, promoting the regeneration of muscle tissue at an accelerated rate. The dosage of **200-400 mcg per session** allows for significant anabolic effects without overwhelming the body's natural processes. On the other hand, **GHK-Cu** serves a dual purpose by not only enhancing muscle repair but

Chapter 5:
Muscle Building with Peptides

also by stimulating collagen and elastin production, vital for maintaining the structural integrity and elasticity of muscle tissues. A weekly dosage of **2-10 mg** strikes the right balance between efficacy and safety, ensuring that the body's healing processes are supported without causing adverse effects.

The concept of **peptide cycling** is particularly important in the context of high-intensity training. By implementing cycles of intensive peptide therapy followed by rest periods, you can maximize the benefits of peptides without risking desensitization. This approach ensures that recovery processes are optimized over time, leading to sustained improvements in performance and resilience. Cycling also allows the body to reset, ensuring that each peptide's effects remain potent and effective over successive cycles.

In addition to peptide therapy, supporting recovery with adequate **nutrition, hydration, and sleep** is essential. A diet rich in anti-inflammatory foods and antioxidants can complement peptide therapy by reducing systemic inflammation and oxidative stress, further accelerating recovery. Hydration plays a critical role in maintaining cellular health and facilitating the transport of nutrients and peptides to damaged tissues. Quality sleep, particularly in the hours immediately following intense training, provides the optimal hormonal environment for recovery, with growth hormone and testosterone levels peaking during deep sleep cycles.

Monitoring your body's response to peptide therapy through regular **blood work** and **biomarker tracking** is crucial for adjusting dosages and protocols to your evolving needs. This personalized approach ensures that peptide therapy remains aligned with your specific recovery requirements, optimizing the healing process and enhancing your ability to withstand the rigors of high-intensity training.

By carefully selecting and combining peptides based on their unique properties and recovery goals, and integrating these with supportive lifestyle practices, individuals can significantly improve their recovery times. This holistic strategy not only accelerates healing but also enhances overall athletic performance, ensuring that you can train harder and more frequently, with a reduced risk of injury. Tailoring the combination and dosage of peptides to your unique physiology and recovery needs, guided by the expertise of healthcare professionals, is essential for achieving optimal results. This personalized, strategic approach to peptide therapy and recovery underscores the importance of understanding and respecting the body's natural healing processes, allowing for a more effective and sustainable enhancement of physical performance and well-being.

Chapter 6: Peptides for Cognitive Clarity

In the realm of cognitive enhancement and brain health, peptides have emerged as a potent tool for individuals looking to sharpen their focus, enhance memory retention, and alleviate stress. These small yet powerful molecules work by mimicking the body's natural signaling processes, thereby influencing a variety of physiological functions related to cognitive performance. One of the primary mechanisms through which peptides exert their effects is by crossing the blood-brain barrier, a critical factor in their ability to directly impact brain function. This ability sets them apart from many other types of supplements and medications that struggle to reach the brain in significant amounts.

Peptides like Semax and Selank have garnered attention for their neuroprotective properties and their potential to improve cognitive functions. Semax, in particular, is known to stimulate the release of brain-derived neurotrophic factor (BDNF), a protein that plays a key role in the growth, maintenance, and survival of neurons. By promoting the activity of BDNF, Semax can enhance learning, memory, and neural plasticity, making it an invaluable asset for anyone looking to boost their mental acuity. Additionally, Semax has been observed to increase attention and memory consolidation, making it particularly useful for students and professionals alike who require extended periods of focus and concentration.

Selank, on the other hand, is praised for its anxiolytic effects, reducing anxiety without the sedative side effects commonly associated with traditional anxiety medications. Its mechanism involves modulating the expression of brain-derived neurotrophic factor (BDNF) and affecting the balance of T helper cell cytokines, contributing to its anti-anxiety and neuroprotective effects. This makes Selank an excellent choice for individuals seeking to manage stress levels while enhancing cognitive clarity and emotional stability.

The synergy between these peptides can provide a comprehensive approach to cognitive enhancement, addressing both the physiological and psychological aspects of brain health. For instance, combining the focus-enhancing properties of Semax with the calming effects of Selank could potentially offer a balanced strategy for achieving heightened mental clarity and reduced stress, thereby facilitating a more productive and focused state of mind.

Moreover, the exploration of peptides for cognitive clarity extends beyond these two compounds. Researchers are continuously investigating the effects of various peptides on brain health, including their potential to mitigate the symptoms of cognitive decline associated with aging and neurodegenerative diseases. This research holds promise for the development of peptide-based protocols tailored to the specific needs of individuals, ranging from enhanced cognitive function in healthy adults to therapeutic applications in patients with cognitive impairments.

As we delve deeper into the specifics of peptide use for cognitive enhancement, it's important to consider the individual variability in response to these compounds. Factors such as dosage, method of administration, and the user's baseline cognitive function can all influence the efficacy of peptide therapy. Therefore, a personalized approach, ideally under the guidance of a healthcare professional experienced in peptide therapy, is crucial for maximizing the benefits while minimizing potential risks. This personalized strategy ensures that each individual can optimize their cognitive function in a safe and effective manner, leveraging the unique properties of peptides to achieve their mental performance goals.

Chapter 6:
Peptides for Cognitive Clarity

Given the intricate nature of peptide therapy for cognitive enhancement, understanding the specific actions and benefits of additional peptides becomes paramount. Dihexa, for instance, is a peptide known for its potent neurogenic properties, significantly outperforming traditional neurotrophic factors like BDNF in promoting synaptic connections and neuronal dendrite growth. Its ability to facilitate the formation of new neural pathways makes Dihexa an exceptional candidate for not only enhancing cognitive function but also for potential therapeutic applications in conditions such as Alzheimer's disease and other forms of dementia. The enhancement of synaptic plasticity by Dihexa implies that it can significantly contribute to learning and memory processes, making it a valuable addition to cognitive enhancement protocols.

Cerebrolysin, another peptide with a broad spectrum of neurotrophic effects, has shown promise in improving cognitive function in patients with neurodegenerative diseases. Its composition, containing free amino acids and short peptides, provides a multifaceted approach to brain health, supporting neuronal growth and reducing neuroinflammatory processes. The neuroprotective properties of Cerebrolysin, alongside its ability to enhance cognitive performance, underscore its potential as a powerful tool in combating cognitive decline and supporting overall brain health.

The integration of peptides such as Dihexa and Cerebrolysin into cognitive enhancement protocols highlights the importance of a targeted approach to peptide therapy. Tailoring peptide selection and dosage to the individual's specific cognitive needs and health status is crucial for achieving optimal results. This requires a comprehensive assessment of cognitive function, alongside regular monitoring to adjust the protocol as necessary. It's also essential to consider the synergistic effects of combining different peptides, as certain combinations may enhance the overall efficacy of the therapy.

Furthermore, the role of lifestyle factors in maximizing the benefits of peptide therapy for cognitive enhancement cannot be overstated. Adequate sleep, a balanced diet rich in omega-3 fatty acids and antioxidants, regular physical exercise, and stress management techniques are all foundational elements that support cognitive health. When combined with peptide therapy, these lifestyle adjustments can significantly amplify the benefits, leading to improved cognitive function, enhanced memory, and reduced mental fatigue.

The evolving landscape of peptide research offers exciting prospects for the future of cognitive enhancement. As our understanding of the brain's biochemistry deepens, the potential for developing more targeted and effective peptide-based therapies continues to expand. This progress promises not only to enhance cognitive function in healthy individuals but also to offer new avenues for the treatment of cognitive disorders, ultimately contributing to the betterment of brain health and cognitive well-being.

Navigating the complexities of peptide therapy for cognitive enhancement requires a careful and informed approach. By staying abreast of the latest research, consulting with knowledgeable healthcare professionals, and adopting a holistic approach to health and wellness, individuals can effectively harness the power of peptides to achieve their cognitive performance goals. The promise of peptides in enhancing cognitive clarity, memory, focus, and overall brain health is an exciting frontier in the realm of biohacking and personal health optimization, offering a glimpse into the future of personalized medicine and cognitive care.

6.1: ENHANCING FOCUS AND MEMORY WITH PEPTIDES

The action of peptides in enhancing focus and memory is a fascinating intersection of neurochemistry and biohacking, offering a promising avenue for individuals seeking to optimize cognitive functions. Peptides, by virtue of their ability to penetrate the blood-brain barrier, can directly influence brain activity, modulating neurotransmitter levels, and enhancing neural connectivity. This direct interaction with the brain's biochemistry allows for a targeted approach to improving cognitive functions, including memory retention and focus. The mechanism through which peptides facilitate these improvements involves the modulation of various signaling pathways that govern synaptic plasticity, the foundation of learning and memory. For instance, peptides can stimulate the production or release of neurotransmitters such as acetylcholine, a critical component in the process of forming new memories and enhancing attention.

Moreover, certain peptides have been shown to upregulate the expression of brain-derived neurotrophic factor (BDNF), a protein that plays a vital role in the survival and growth of neurons. BDNF not only supports the existing neural infrastructure but also promotes the growth of new neurons and synapses, a process known as neurogenesis. This effect on BDNF levels is particularly beneficial for long-term cognitive health, offering protection against cognitive decline and supporting enhanced neural function. In addition to their neurogenic and neuroprotective effects, peptides can also exert antioxidative and anti-inflammatory actions within the brain, further contributing to an optimal cognitive environment. By reducing oxidative stress and inflammation, peptides help preserve the integrity of neural tissues, ensuring that the signaling processes critical for memory and focus remain efficient.

Chapter 6:
Peptides for Cognitive Clarity

When considering the incorporation of peptides into a cognitive enhancement strategy, it is essential to approach with precision, tailoring peptide selection and dosing to individual needs and goals. For enhancing focus and memory, peptides such as Noopept, which is known for its ability to improve synaptic transmission in the brain, and Piracetam, which enhances cellular membrane fluidity and thus improves communication between neurons, are often recommended. Starting with lower doses and gradually adjusting based on response allows for the identification of an effective yet safe dosing regimen. It is also crucial to monitor cognitive function over time to assess the efficacy of the peptide protocol and make necessary adjustments.

Combining peptide therapy with other cognitive enhancement strategies, such as nutrient optimization, adequate sleep, and cognitive training exercises, can amplify the benefits achieved through peptides alone. Nutrients like omega-3 fatty acids, antioxidants, and B-vitamins support brain health and function, while sleep plays a critical role in memory consolidation and cognitive recovery. Cognitive training exercises, such as memory games and problem-solving tasks, can further enhance the brain's plasticity, making it more receptive to the benefits offered by peptides.

In summary, the use of peptides for improving focus and memory represents a cutting-edge approach to cognitive enhancement, grounded in an understanding of neurobiology and personalized health care. By carefully selecting and dosing peptides, individuals can harness their potential to support and enhance cognitive functions, achieving greater mental clarity, improved memory retention, and a heightened ability to concentrate. As research in this area continues to evolve, so too will the strategies for effectively utilizing peptides to achieve cognitive optimization, offering exciting possibilities for those committed to personal health and performance enhancement.

6.2: ESSENTIAL PEPTIDES FOR COGNITIVE FUNCTION

Dihexa and Cerebrolysin stand out as potent peptides for cognitive enhancement, each with unique mechanisms of action that support brain health and function. Dihexa, for instance, is a peptide known for its ability to significantly increase synaptic connections, which are the points of communication between neurons. This enhancement in neural connectivity is crucial for learning, memory formation, and overall cognitive agility. The remarkable aspect of Dihexa is its potency; it has been shown in studies to be many times more effective than BDNF (Brain-Derived Neurotrophic Factor) in promoting neural connections, making it a powerful candidate for addressing cognitive decline and improving mental acuity. Its application could be particularly beneficial for individuals experiencing age-related cognitive slowdown or those looking to boost their cognitive reserves against neurodegenerative conditions.

Cerebrolysin, on the other hand, is a peptide with a broader range of neuroprotective properties. It is composed of low-molecular-weight peptides and amino acids that mimic the action of endogenous neurotrophic factors, promoting brain health and function. Cerebrolysin has been observed to enhance cognitive function, memory, and has neuroregenerative properties that make it valuable in the treatment and management of Alzheimer's disease, stroke recovery, and traumatic brain injuries. Its multifaceted approach to brain health includes improving synaptic density, enhancing neuroplasticity, and reducing amyloid-beta plaques, which are associated with cognitive decline in Alzheimer's disease.

For individuals seeking to incorporate these peptides into their cognitive health regimen, it is crucial to understand the recommended protocols for their use. Dihexa, due to its potent nature, requires precise dosing that should be tailored to the individual's specific health profile and cognitive goals. Starting with a lower dose and gradually increasing based on tolerance and observed effects can be a prudent approach. Monitoring cognitive function through standardized tests or assessments can help gauge the effectiveness of the regimen and guide dosage adjustments.

Cerebrolysin administration typically involves a course of treatment that may vary in duration and dosage, depending on the individual's condition and therapeutic goals. It is often administered intravenously or intramuscularly, which necessitates professional oversight by a healthcare provider. The course of treatment with Cerebrolysin may range from a few weeks to several months, and its effects on cognitive function can be enhanced by combining it with lifestyle factors such as a balanced diet, regular physical exercise, and cognitive training exercises.

Both Dihexa and Cerebrolysin represent promising avenues for cognitive enhancement and neuroprotection. However, their use must be approached with caution, under the guidance of a healthcare professional experienced in peptide therapies. This ensures not only the safety and efficacy of the treatment but also allows for personalized protocols that can maximize cognitive benefits while minimizing potential risks. As research continues to unveil the full potential of these and other cognitive-enhancing peptides, they remain valuable tools in the quest for optimal brain health and function.

6.2.1: Semax and Selank for Memory and Mood

Chapter 6:
Peptides for Cognitive Clarity

Semax and Selank, two peptides with significant neurogenic and neuroprotective properties, have garnered attention for their potential to enhance cognitive function, particularly in memory retention, stress reduction, and mood improvement. Semax, primarily known for its cognitive-enhancing effects, operates by stimulating the release and expression of brain-derived neurotrophic factor (BDNF), a protein integral to the growth and differentiation of new neurons and synapses. This action not only aids in the improvement of cognitive processes such as learning and memory but also plays a crucial role in the brain's adaptability to various stressors, thereby enhancing its resilience. The recommended starting dosage for Semax is typically low, with incremental adjustments based on individual response and tolerance. The administration route is usually nasal, which facilitates direct access to the brain, enhancing its bioavailability and efficacy.

Selank, on the other hand, is a synthetic analogue of the human tetrapeptide tuftsin. It is primarily recognized for its anxiolytic properties without the sedative effects commonly associated with traditional anxiolytics. By modulating the expression of Interleukin-6 (IL-6) and influencing the balance of T helper cell cytokines, Selank can effectively reduce anxiety levels and improve emotional stability. This modulation of immune response also contributes to its cognitive-enhancing effects, particularly in memory and attention. Selank's administration, similar to Semax, is typically through a nasal spray, offering a convenient and efficient method of delivery. The dosage of Selank should be tailored to the individual, starting with the lower end of the recommended spectrum and adjusting based on personal tolerance and the specific cognitive or mood-related goals aimed to be achieved.

The synergistic use of Semax and Selank presents a compelling approach for those seeking to optimize cognitive functions and emotional well-being. When considering incorporating these peptides into a cognitive health regimen, it is crucial to understand the importance of starting with lower doses to assess individual tolerance. Additionally, monitoring one's response over time allows for the adjustment of dosages to optimize the therapeutic benefits while minimizing potential side effects. The combination of Semax and Selank, with their distinct yet complementary mechanisms of action, offers a multifaceted approach to enhancing cognitive function, reducing stress, and improving mood. This dual peptide protocol can be particularly beneficial for individuals facing high cognitive demands or those seeking to improve their mental health and resilience.

Incorporating lifestyle factors such as a balanced diet rich in omega-3 fatty acids, regular physical exercise, and adequate sleep further supports the effectiveness of Semax and Selank in cognitive and emotional health. The neuroprotective and neurogenic effects of these peptides, combined with a healthy lifestyle, provide a solid foundation for achieving and maintaining optimal cognitive function and emotional well-being. It is also imperative to consult with a healthcare professional experienced in peptide therapies to ensure the safe and effective use of Semax and Selank, tailored to one's health status and cognitive enhancement goals. This personalized approach maximizes the potential benefits of these peptides, contributing to the individual's overall cognitive health strategy.

6.2.2: Dihexa and Cerebrolysin

Dihexa and Cerebrolysin, as cognitive enhancers, offer a beacon of hope for those grappling with cognitive decline or seeking to preemptively bolster their cognitive resilience. The nuanced understanding of these peptides and their application in cognitive health protocols requires a meticulous approach, blending scientific insight with practical application strategies. Dihexa, with its unparalleled ability to foster neural connection, stands at the forefront of neuroregenerative research. Its mechanism, centered around augmenting the synaptic connections, lays the groundwork for profound improvements in cognitive functions such as memory, learning, and overall brain agility. The specificity of Dihexa's action on neural connectivity not only underscores its potential as a therapeutic agent for neurodegenerative diseases but also as a preventive measure for age-related cognitive decline. The administration of Dihexa necessitates a personalized approach, taking into account the individual's cognitive baseline and health profile. This tailored strategy ensures the maximization of Dihexa's cognitive benefits while mitigating potential risks associated with its potency.

Cerebrolysin, with its broad spectrum of neuroprotective properties, complements Dihexa's targeted approach by offering a holistic enhancement of brain health. Its composition, a complex blend of peptides and amino acids, mirrors the body's own neurotrophic factors, thereby supporting brain function on multiple fronts. From enhancing synaptic density and neuroplasticity to reducing neuroinflammatory markers, Cerebrolysin addresses the multifaceted challenges of cognitive impairment and neurodegeneration. Its role in cognitive rehabilitation, particularly post-stroke and in traumatic brain injuries, highlights its capacity to not only protect but also repair and regenerate brain tissue. The administration of Cerebrolysin, often through intravenous or intramuscular routes, requires professional oversight, ensuring that the treatment protocol is aligned with the individual's specific therapeutic needs and goals.

The integration of Dihexa and Cerebrolysin into a cognitive health regimen opens new avenues for cognitive enhancement and neuroprotection. However, this integration is not without its complexities. The potent nature of these peptides demands a cautious approach, emphasizing the importance of starting with lower doses and closely monitoring cognitive function to guide dosage adjustments. This vigilant strategy, coupled with regular health assessments, ensures that the therapeutic potential of Dihexa and Cerebrolysin is fully realized while minimizing adverse effects. Furthermore, the synergy between these peptides and lifestyle modifications cannot be overstated. A holistic approach, incorporating a balanced diet, regular physical activity, and cognitive exercises, amplifies the cognitive benefits of Dihexa and Cerebrolysin, creating a robust framework for cognitive health optimization.

The evolving landscape of peptide therapy, with Dihexa and Cerebrolysin at its helm, represents a promising frontier in the quest for cognitive vitality and longevity. As research continues to unravel the intricacies of these peptides and their mechanisms of action, their role in personalized medicine becomes increasingly significant. This personalized approach, grounded in a deep understanding of individual health profiles and cognitive goals, paves the way for targeted, effective interventions. The potential of Dihexa and Cerebrolysin to transform cognitive health practices is immense, offering a glimpse into a future where cognitive decline is no longer an inevitable consequence of aging but a challenge that can be met with innovative, evidence-based solutions.

6.3: Long-Term Mental Health Protocols

For long-term mental health support, the strategic implementation of peptide protocols can play a pivotal role in sustaining cognitive clarity and resilience. The foundation of these protocols lies in the cyclic and targeted use of peptides, tailored to individual needs and adjusted over time as those needs evolve. It's crucial to establish a baseline of cognitive function, utilizing cognitive assessments and biomarker analysis, to personalize peptide protocols effectively. This approach allows for the optimization of peptide types, dosages, and cycles, ensuring they align with the ongoing mental health objectives of the individual.

Semax and **Selank**, for instance, can be alternated in cycles to manage stress levels and enhance cognitive function without leading to receptor desensitization. A typical cycle might involve using Semax for a period of two to three weeks, followed by a similar duration of Selank use. This rotation helps maintain the efficacy of each peptide while minimizing potential side effects. The dosages should start on the lower end of the spectrum, gradually increasing based on individual tolerance and response. Monitoring through regular cognitive assessments aids in fine-tuning the protocol to the individual's evolving mental health landscape.

Incorporating **Dihexa** and **Cerebrolysin** into a long-term mental health strategy requires careful consideration of their potent effects on neural connectivity and neuroprotection. Given their powerful impact, these peptides are best utilized in shorter cycles, interspersed with periods of rest or the use of other, less potent peptides. For Dihexa, a cycle might last one to two weeks, followed by a two to four-week break or a cycle of a different peptide aimed at supporting neural health, such as Cerebrolysin. The key is to prevent overstimulation of neural pathways, allowing the brain to integrate the enhancements without becoming overwhelmed.

The role of lifestyle factors in conjunction with peptide protocols cannot be overstated. A diet rich in omega-3 fatty acids, antioxidants, and neuroprotective nutrients supports the structural and functional optimization provided by peptides. Regular physical exercise, particularly aerobic activities, enhances cerebral blood flow and neurogenesis, creating an ideal environment for peptides to exert their benefits. Additionally, practices such as mindfulness meditation, cognitive exercises, and adequate sleep hygiene enhance the stress-reducing and cognitive-enhancing effects of peptides, creating a comprehensive approach to long-term mental health.

Regular consultations with healthcare professionals experienced in peptide therapy are essential for the safe and effective implementation of these protocols. These professionals can provide guidance on adjusting dosages, cycling strategies, and integrating lifestyle adjustments to maximize the benefits of peptide use for cognitive health. They also play a crucial role in monitoring the individual's response to the therapy, making necessary adjustments based on feedback from cognitive assessments and biomarker analysis.

Adherence to these protocols, combined with a commitment to a supportive lifestyle and regular monitoring, lays the groundwork for sustained mental vitality and resilience. It's a dynamic process, requiring adjustments and refinements as the individual's needs and responses evolve. The ultimate goal is to support a state of mental well-being that fosters clarity, focus, and resilience, enabling individuals to navigate the complexities of life with enhanced cognitive capacity and emotional stability.

Chapter 7: Immune System Boost

7.1: Peptides in Immune Health

Peptides play a crucial role in modulating the immune system, enhancing its ability to fend off pathogens and reduce the incidence of infections. These small yet powerful molecules can influence various aspects of the immune response, from increasing the production of antibodies to enhancing the activity of natural killer (NK) cells, which are pivotal in the body's defense against viruses and tumor cells. For individuals seeking to bolster their immune health, understanding the specific peptides that have been shown to have immunomodulatory effects is essential. Thymosin Alpha-1, for instance, has been extensively studied for its ability to augment immune function. It achieves this by promoting the maturation of T cells from the thymus, which are essential for adaptive immunity, and by stimulating the production of cytokines, which are critical for immune communication and response coordination. This peptide is particularly beneficial for individuals with chronic infections or those who are immunocompromised, as it can significantly enhance the body's ability to fight off pathogens.

Another peptide, LL-37, plays a multifaceted role in the immune system. It not only exhibits direct antimicrobial activity against a wide range of pathogens, including bacteria, viruses, and fungi, but also modulates the immune response, reducing excessive inflammation which can be detrimental to tissue health and immune function. LL-37's ability to disrupt biofilms, which are protective layers formed by bacteria to shield themselves from antibiotics and the immune system, makes it a valuable asset in treating chronic bacterial infections that are resistant to conventional treatments.

KPV, a shorter segment of the larger peptide LL-37, retains anti-inflammatory and antimicrobial properties. It has been shown to reduce inflammation and microbial growth in various models, making it a promising candidate for managing conditions characterized by inflammation and infection. The application of KPV could be particularly useful in dermatological conditions, where inflammation and microbial infections often coexist.

For individuals interested in utilizing peptides to support their immune health, it is important to consider the method of administration, as this can influence the efficacy of the peptide. While some peptides like Thymosin Alpha-1 are commonly administered through subcutaneous injections, others may be available in topical forms for conditions such as skin infections or inflammations. The dosage and frequency of administration will vary depending on the peptide, the specific health condition being addressed, and the individual's overall health profile. Consulting with a healthcare professional experienced in peptide therapy is crucial to determine the most appropriate peptide, dosage, and administration route for one's specific needs.

In addition to peptide therapy, supporting immune health involves a holistic approach that includes a balanced diet rich in nutrients that support immune function, regular physical activity, adequate sleep, and stress management. These lifestyle factors can significantly influence the effectiveness of peptide therapy in boosting immune health. For instance, a diet lacking in essential vitamins and minerals can impair the body's immune response, while chronic stress has been shown to suppress immune function. Therefore, integrating peptide therapy with a healthy lifestyle can synergize to enhance immune resilience and reduce the risk of infections.

It is also important to monitor the response to peptide therapy through regular health checkups and blood tests, as this can provide insights into the effectiveness of the treatment and whether any adjustments are needed. Biomarkers of immune function, such as levels of specific antibodies or the activity of NK cells, can offer valuable information on how well the immune system is responding to the peptides.

In summary, peptides offer a promising avenue for enhancing immune health and preventing infections. By selectively modulating the immune response, peptides such as Thymosin Alpha-1, LL-37, and KPV can enhance the body's natural defenses, reduce inflammation, and combat pathogens. However, the success of peptide therapy in boosting immune health depends on a comprehensive approach that includes appropriate peptide selection, method of administration, dosage, and integration with healthy lifestyle practices. With the guidance of healthcare professionals and a commitment to overall well-being, individuals can harness the power of peptides to support a robust and resilient immune system.

7.2: BEST PEPTIDE COMBINATIONS FOR IMMUNITY

Building on the foundation of understanding individual peptides' roles in enhancing immune health, it's crucial to explore how combining these peptides can synergize to create more robust immune support. The strategic pairing of peptides can address various facets of the immune system, from enhancing antibody production to modulating the inflammatory response, thereby offering a comprehensive approach to immune resilience. A notable combination involves the use of **Thymosin Beta-4 (TB-500)** with **Thymosin Alpha-1**. Thymosin Beta-4 is primarily recognized for its healing properties, significantly improving wound healing and tissue repair. When used in conjunction with Thymosin Alpha-1, known for its immune-enhancing capabilities, the duo works to not only bolster the body's defense mechanisms but also to facilitate rapid recovery from injuries, thereby maintaining the integrity of the body's first line of defense against pathogens.

Another powerful combination is **CJC-1295** and **Ipamorelin**, peptides that, when used together, stimulate the release of growth hormone (GH) from the pituitary gland. This increase in GH levels has a cascading effect on immune health, as GH indirectly supports the function of T-cells and NK cells. The synergy between CJC-1295 and Ipamorelin can thus enhance the body's natural immune response while also providing the added benefits of improved sleep and recovery, factors that are intrinsically linked to immune function.

For targeting inflammation, a common underlying factor in many chronic diseases and immune dysfunctions, the combination of **BPC-157** and **LL-37** offers a promising solution. BPC-157 is known for its potent anti-inflammatory properties and its ability to promote healing in damaged tissues, including the gut lining, where a significant portion of the immune system resides. LL-37, with its antimicrobial and immune-modulating effects, complements BPC-157 by providing a protective barrier against pathogens and reducing harmful inflammation. This combination not only addresses the symptoms of immune dysregulation but also targets the root causes, promoting long-term immune health.

In the realm of autoimmune conditions, where the immune system mistakenly attacks the body's own tissues, the use of **VIP (Vasoactive Intestinal Peptide)** in combination with **Thymosin Alpha-1** has shown promise. VIP has a regulatory effect on the immune system, helping to balance immune response and prevent overactivation that can lead to tissue damage. Thymosin Alpha-1 supports this by enhancing the body's ability to distinguish between self and non-self, reducing the likelihood of autoimmune reactions. This combination is particularly beneficial for those with autoimmune challenges, offering a pathway to modulate the immune response and promote tolerance.

It's important to note that while these combinations offer significant benefits, the approach to peptide therapy should always be personalized. Factors such as individual health status, existing conditions, and specific immune health goals must be considered when selecting peptide combinations. Consulting with a healthcare professional experienced in peptide therapy is essential to tailor the protocol to the individual's needs, ensuring optimal safety and efficacy. Additionally, integrating peptide therapy with a healthy lifestyle, including a nutrient-rich diet, regular exercise, adequate sleep, and stress management, will further enhance the immune-boosting effects of peptides. By adopting a comprehensive and personalized approach to peptide therapy, individuals can significantly improve their immune resilience, equipping their bodies to better withstand and recover from health challenges.

7.2.1: Thymosin Alpha-1 for Immune Modulation

Thymosin Alpha-1, a key peptide in the realm of immune modulation, emerges as a potent ally in the fight against infections and in bolstering the body's defense mechanisms. This peptide, a small protein fragment, plays a pivotal role in the regulation of immune responses, enhancing the activity of T-cells which are crucial for the body's ability to combat pathogens and diseases. Its mechanism of action involves the modulation of various immune cells, promoting a shift towards an environment that is not only conducive to fighting off infections but also to preventing them.

Chapter 7:
Immune System Boost

The administration of Thymosin Alpha-1 has been shown to significantly augment the function of immune cells, particularly in individuals whose immune system is compromised or in need of support due to stress, aging, or other factors that can lead to a decrease in immune function. The peptide achieves this by influencing the differentiation and maturation of T-cells from the thymus, an organ that is central to the development of a robust immune system but which naturally diminishes in function with age. By enhancing the activity of T-cells, Thymosin Alpha-1 aids in the establishment of a more vigilant immune surveillance system, capable of swiftly identifying and responding to the presence of pathogens.

For individuals seeking to incorporate Thymosin Alpha-1 into their immune support regimen, it is imperative to understand the recommended protocols for its use. Typically, the peptide is administered via subcutaneous injection, allowing for direct absorption into the bloodstream and ensuring that it reaches the cells of the immune system efficiently. The dosage and frequency of administration can vary depending on individual needs and goals, with a common regimen involving low-dose injections administered several times a week over a course of treatment that may span several weeks to months, depending on the desired outcome and under the guidance of a healthcare professional.

It is also crucial to source Thymosin Alpha-1 from reputable suppliers to ensure the purity and potency of the peptide, as the market for peptides can vary widely in terms of quality and safety standards. Verification of the product through third-party testing and consultation with healthcare providers familiar with peptide therapies can help ensure that individuals receive a product that is safe and effective for use.

In addition to its direct effects on immune cell function, Thymosin Alpha-1 has been observed to have synergistic effects when used in combination with other peptides and treatments aimed at enhancing immune health. For instance, combining Thymosin Alpha-1 with peptides that promote gut health or reduce inflammation can lead to a more comprehensive approach to immune support, addressing multiple pathways through which the immune system can be bolstered.

The potential of Thymosin Alpha-1 to modulate the immune system and enhance the body's ability to fight infections makes it a valuable tool in the arsenal of peptides for health optimization. Its use, guided by a thorough understanding of its mechanisms, proper administration, and adherence to safety protocols, can offer significant benefits for individuals looking to support their immune health and resilience against infections.

7.2.2: LL-37 and KPV for Infection Defense

LL-37 and KPV stand as formidable allies in the defense against bacterial and viral infections, leveraging their unique antimicrobial and immunomodulatory properties to fortify the body's natural defenses. LL-37, a cathelicidin-derived peptide, operates through a multifaceted mechanism, directly targeting and disrupting the membranes of invading pathogens, thus neutralizing their ability to proliferate and cause disease. Its role extends beyond mere antimicrobial action; LL-37 also modulates the immune response, enhancing the body's ability to mount an effective defense against infections while mitigating excessive inflammation that can lead to tissue damage. This dual functionality not only makes LL-37 a critical component in the fight against pathogens but also positions it as a therapeutic agent in managing inflammatory conditions and wound healing processes.

KPV, a tripeptide with a sequence of Lysine-Proline-Valine, exhibits potent anti-inflammatory and immunomodulatory effects, making it an invaluable peptide in the management and prevention of infections. By downregulating inflammatory cytokines and inhibiting the activation of NF-kB, a protein complex that plays a key role in regulating the immune response to infection, KPV helps to maintain a balanced immune environment. This equilibrium is crucial for preventing the overactivation of immune responses that can lead to autoimmune diseases and allergies, as well as for ensuring an adequate defense against pathogens.

The administration of LL-37 and KPV, either as standalone treatments or in combination, offers a strategic approach to enhancing the body's infection defense mechanisms. For individuals seeking to incorporate these peptides into their immune support regimen, it is essential to understand the specific protocols for their use. The dosage of LL-37 typically ranges from 100 to 300 micrograms per kilogram of body weight, administered via subcutaneous injection. This dosage can be adjusted based on individual needs, the nature of the infection, and the guidance of a healthcare professional. KPV, on the other hand, is administered at lower doses, often in the range of 100 to 500 micrograms, due to its potent effects even at minimal concentrations.

When initiating a protocol involving LL-37 and KPV, starting with a lower dose and gradually increasing it allows for the monitoring of tolerance and effectiveness, minimizing the risk of adverse reactions. The frequency of administration is another critical factor, with daily injections being common during the initial phase of treatment, followed by a maintenance phase with less frequent dosing. The duration of the protocol can vary, extending from a few weeks to several months, depending on the therapeutic goals, the severity of the condition being treated, and the individual's response to the peptides.

Sourcing these peptides from reputable suppliers is paramount to ensure their purity, potency, and safety. Engaging with a healthcare provider who has expertise in peptide therapies is equally important, as they can offer personalized advice, adjust treatment protocols as necessary, and monitor progress through regular follow-ups and laboratory tests. This professional guidance is invaluable in maximizing the benefits of LL-37 and KPV while safeguarding against potential risks.

In conclusion, LL-37 and KPV represent powerful tools in the enhancement of the immune system's ability to defend against infections. Their targeted mechanisms of action, combined with the ability to modulate the immune response, make them suitable candidates for individuals looking to bolster their infection defense strategies. With careful consideration of dosing protocols, adherence to safety guidelines, and consultation with healthcare professionals, the use of these peptides can contribute significantly to maintaining and improving health and resilience against pathogens.

7.3: USING PEPTIDES FOR PREVENTATIVE HEALTH

Incorporating peptides as preventative health measures offers a proactive approach to maintaining a robust immune system, crucial for warding off diseases and infections. This strategy involves the use of specific peptides known for their immune-boosting properties, aiming to enhance the body's natural defense mechanisms before the onset of illness. To effectively implement peptides in a preventative regimen, it's essential to understand the types of peptides that serve this purpose, their mechanisms of action, and the optimal protocols for their use.

Peptides such as **Thymosin Beta-4**, **Melanotan II**, and **VIP (Vasoactive Intestinal Peptide)** are among those recognized for their potential in immune system support. Thymosin Beta-4, for instance, plays a pivotal role in tissue repair, inflammation control, and the enhancement of T-cell and dendritic cell response. Melanotan II, while primarily known for its effects on melanogenesis and thus skin protection, also exhibits anti-inflammatory properties. VIP, on the other hand, has shown significant promise in regulating immune responses, protecting against autoimmune diseases, and promoting a balanced immune system.

The mechanism of action for these peptides varies but generally revolves around their ability to modulate immune cell activity, promote the production of antibodies, and facilitate communication between cells of the immune system. This modulation helps the body to maintain a state of readiness against pathogens, effectively reducing the likelihood of infections and enhancing the overall resilience of the immune system.

For individuals considering the inclusion of these peptides in their health regimen, it's advisable to start with a comprehensive health assessment. This evaluation should focus on current immune health status, potential risk factors for immune-related issues, and specific health goals. Based on this assessment, a tailored peptide protocol can be developed, taking into account factors such as dosage, frequency, and duration of peptide use. It's important to note that while peptides can significantly bolster the immune system, they should complement other health practices such as a balanced diet, regular exercise, adequate sleep, and stress management.

The administration of immune-boosting peptides typically involves subcutaneous injections, as this method ensures direct absorption into the bloodstream and immediate availability to immune cells. The dosage and frequency of injections depend on the specific peptide and the individual's health status and goals. For example, a common protocol might involve administering Thymosin Beta-4 at a dose of 2-10 mg per week, divided into two or more injections. These protocols may vary, and it's crucial to consult with a healthcare professional experienced in peptide therapy to determine the most appropriate regimen.

Furthermore, the timing of peptide administration can play a critical role in their effectiveness as preventative measures. Some peptides may be more beneficial when used in anticipation of periods known for increased risk of illness, such as flu season, or during times of heightened stress when the immune system is more vulnerable.

Sourcing peptides from reputable suppliers is another critical aspect of using peptides safely and effectively. Products should be verified for purity and potency, ensuring that they meet the necessary standards for therapeutic use. Additionally, ongoing monitoring and adjustment of peptide protocols are essential, as individual responses to peptide therapy can vary. Regular follow-ups with a healthcare provider allow for the assessment of effectiveness and the making of any necessary adjustments to the protocol.

In summary, the strategic use of peptides as preventative health measures represents a forward-thinking approach to enhancing immune resilience and overall well-being. By carefully selecting peptides known for their immune-supportive properties, adhering to recommended protocols, and integrating these with general health practices, individuals can significantly contribute to the strength and efficiency of their immune system. This proactive stance not only helps in preventing illness but also supports a foundation for long-term health and vitality.

7.3.1: Seasonal Immune Protocols

Seasonal immune protocols are designed to bolster the body's defenses during times when it is most susceptible to illness, such as the flu season or periods of seasonal allergies. These protocols leverage the strategic use of peptides known for their immune-enhancing properties, timed to preemptively counteract the increased risk of infections and diseases prevalent during specific times of the year. The cornerstone of these protocols is the anticipation of the body's needs, adjusting peptide regimens to provide optimal support before the immune system is challenged.

As the seasons change, so do the environmental factors that can impact immune health, including variations in temperature, humidity, and the presence of pathogens. During the fall and winter months, for instance, the prevalence of influenza and the common cold necessitates a proactive approach to immune support. Peptides such as Thymosin Beta-4 and Thymosin Alpha-1 can be particularly beneficial during these times, enhancing T-cell function and supporting the body's natural virus-fighting abilities. A typical protocol might involve the administration of Thymosin Beta-4 at a dosage of 2-10 mg per week, starting several weeks before the season's peak and continuing throughout its duration to maintain heightened immune vigilance.

In contrast, spring often brings an increase in allergic responses due to pollen and other allergens. During this time, peptides that modulate the immune response to reduce inflammation and allergic reactions can be invaluable. For example, administering a peptide like VIP (Vasoactive Intestinal Peptide), which has shown promise in regulating immune responses and protecting against autoimmune diseases, can help mitigate the severity of seasonal allergies. Dosing for VIP might range from 25 to 50 mcg daily, initiated before the onset of the allergy season and adjusted as needed based on symptom severity and individual response.

Moreover, transitioning into summer, the focus shifts towards maintaining overall immune resilience in the face of increased outdoor activities and potential exposure to diverse pathogens. Protocols during this period might emphasize peptides that support general immune health, such as Thymosin Alpha-1, to ensure the immune system remains robust and fully functional. Additionally, peptides that promote skin health and repair, such as GHK-Cu, become relevant as they aid in the recovery from sun exposure and minor skin injuries, common during summer months.

The implementation of seasonal immune protocols requires careful planning and consideration of individual health status, history of seasonal illnesses, and lifestyle factors. It is crucial to begin peptide administration ahead of the season's start to allow the immune system to adapt and respond effectively to the enhanced support. Regular consultation with a healthcare provider experienced in peptide therapy is essential to tailor the protocol to the individual's needs, monitor progress, and make adjustments as necessary.

Adjusting peptide types, dosages, and administration frequency according to seasonal demands enables a dynamic approach to immune support, addressing the body's fluctuating requirements throughout the year. This proactive strategy not only aims to prevent seasonal illnesses but also supports optimal immune function as part of a comprehensive health and wellness regimen. By aligning peptide therapy with the body's natural rhythms and seasonal challenges, individuals can significantly enhance their resilience against infections, allergies, and other immune-related concerns, maintaining health and vitality regardless of the season.

7.3.2: Guidelines for Chronic Immune Support

For individuals facing chronic immune challenges or prolonged exposure to immune-compromising conditions, a strategic and sustained approach to peptide therapy can serve as a cornerstone of their health maintenance regimen. This approach necessitates a nuanced understanding of how peptides function within the immune system and the ways in which they can be leveraged to provide ongoing support. The goal is to create an environment where the immune system operates at an optimal level, despite the challenges posed by chronic conditions or environmental factors.

The selection of peptides for chronic immune support should be guided by their specific properties and the mechanisms through which they influence immune function. Peptides such as Thymosin Alpha-1 have demonstrated significant efficacy in modulating immune responses, making them prime candidates for long-term immune support. However, the effectiveness of peptide therapy in a chronic context relies not only on the choice of peptide but also on the regimen followed.

A continuous or long-term protocol must be carefully calibrated to maintain its effectiveness over time. This involves not only the dosage and frequency of administration but also the consideration of potential adaptations to the protocol to prevent the immune system from becoming unresponsive to the therapy. For instance, periodic evaluations may reveal the need for adjusting dosages or introducing "rest" periods during which peptide administration is paused to allow the immune system to reset.

Moreover, the integration of peptide therapy into a broader health maintenance strategy is crucial for individuals with chronic immune challenges. This includes

Chapter 7:
Immune System Boost

nutritional support tailored to bolster immune function, such as diets rich in antioxidants and anti-inflammatory foods, as well as lifestyle modifications like stress reduction techniques and adequate sleep, which have been shown to significantly impact immune health.

The administration of peptides in the context of chronic immune support should also be approached with an eye toward safety and sustainability. Subcutaneous injections are commonly used for their efficiency in delivering peptides directly into the bloodstream, but the technique must be performed correctly to minimize risks. Education on proper injection practices, awareness of potential side effects, and regular monitoring by healthcare professionals are essential components of a safe long-term peptide therapy regimen.

Collaboration with a healthcare provider experienced in peptide therapy is invaluable for individuals embarking on a long-term peptide-based immune support strategy. This partnership enables the customization of the peptide protocol to the individual's unique health profile, ongoing assessment of the therapy's effectiveness, and adjustments as necessary based on the body's response.

Given the dynamic nature of the immune system and the complexities involved in managing chronic immune challenges, the use of peptides as a preventative health measure offers a promising avenue for enhancing quality of life and health outcomes. Through careful selection, meticulous protocol design, and integration with broader health strategies, peptides can provide significant support to those seeking to maintain immune resilience over the long term. This proactive approach, underpinned by a commitment to regular evaluation and adjustment, ensures that peptide therapy remains a viable and effective component of chronic immune support, tailored to meet the evolving needs of the individual without the need for closing remarks or summary statements to underscore its potential benefits.

Chapter 8: Anti-Aging and Cellular Regeneration

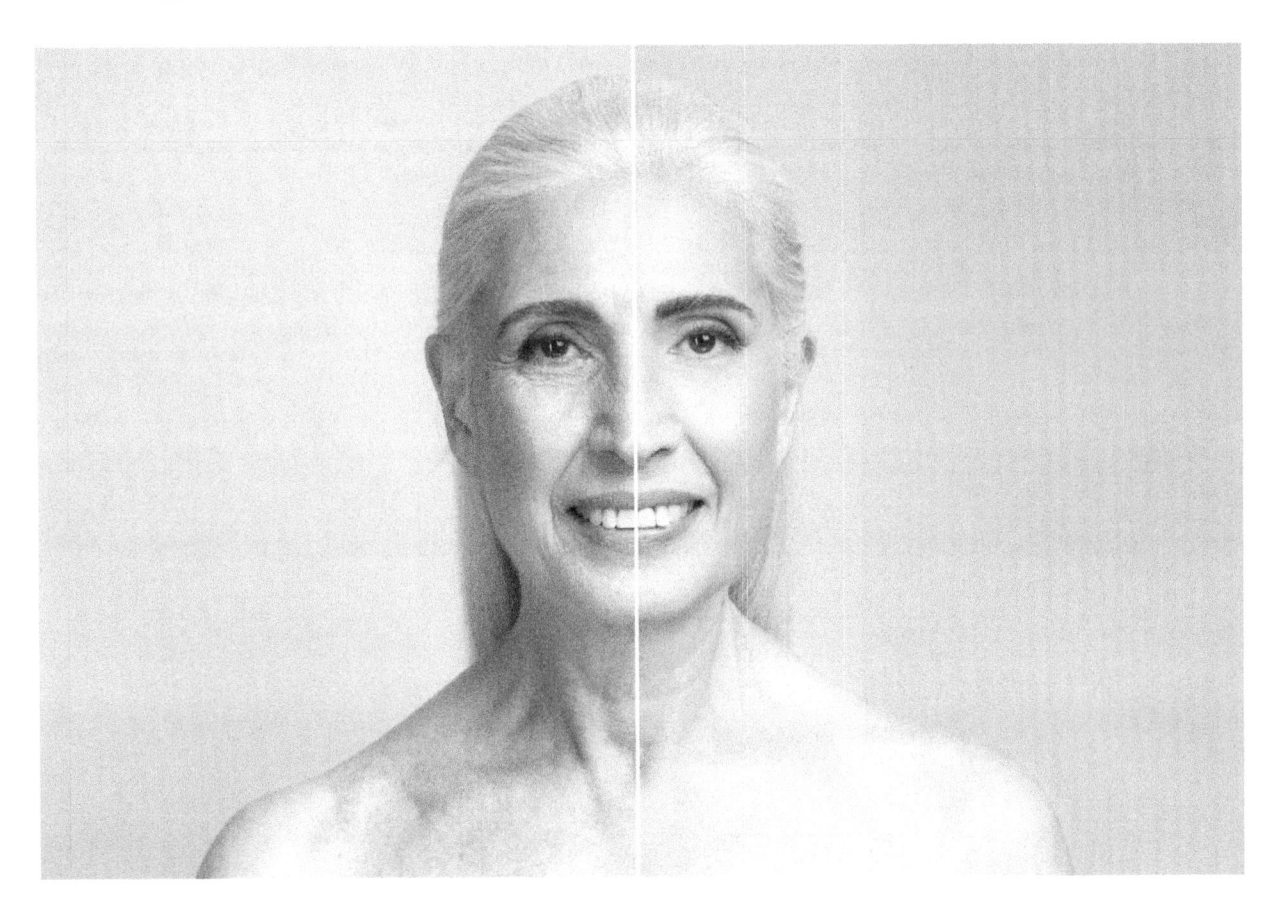

8.1: Peptides and Cellular Longevity

Peptides, due to their fundamental role in cellular processes, have emerged as a cornerstone in the quest for cellular longevity and anti-aging. These small chains of amino acids are not just building blocks but are active participants in various cellular functions, including repair, replication, and signaling. Their impact on cellular longevity can be attributed to their ability to influence gene expression, modulate immune responses, and enhance the repair mechanisms of damaged DNA. For instance, peptides like Epithalon have been shown to stimulate the production of telomerase, an enzyme that plays a crucial role in the maintenance and repair of telomeres. Telomeres, the protective caps at the ends of chromosomes, naturally shorten with each cell division, leading to cellular aging and eventual cell death. By promoting telomerase activity, peptides can effectively slow down the telomere shortening process, thereby extending the lifespan of cells and contributing to overall cellular rejuvenation.

Moreover, peptides such as GHK-Cu not only promote skin and tissue repair but also exhibit potent anti-inflammatory and antioxidant properties. These effects are critical in combating the oxidative stress and chronic inflammation associated with aging. Oxidative stress, resulting from an imbalance between free radicals and antioxidants in the body, damages cells and tissues, accelerating the aging process. By neutralizing free radicals and reducing inflammation, peptides help preserve cellular integrity and function, further supporting longevity.

The administration of these peptides, typically through subcutaneous injections, allows for direct absorption into the bloodstream, ensuring that they reach target cells and tissues effectively. The dosage, frequency, and duration of peptide therapy must be carefully calibrated based on individual health status, goals, and under the guidance of a healthcare professional experienced in peptide therapies. It is also essential to source peptides from reputable suppliers to ensure their purity, potency, and safety.

In addition to direct interventions, lifestyle factors play a significant role in maximizing the benefits of peptide therapy for cellular longevity. A balanced diet rich in antioxidants, regular physical activity, adequate sleep, and stress management techniques are foundational elements that support the efficacy of peptides in promoting cellular health and longevity. These lifestyle adjustments not only enhance the body's natural repair mechanisms but also optimize the internal environment, making it more conducive for peptides to exert their beneficial effects.

Furthermore, ongoing research and advancements in peptide science continue to unveil new peptides with potential anti-aging benefits, highlighting the dynamic and evolving nature of this field. As our understanding of the molecular mechanisms underlying aging advances, so too will the development of targeted peptide therapies aimed at extending cellular lifespan and improving the quality of life.

Incorporating peptides into a comprehensive anti-aging strategy requires a personalized approach, taking into account the unique physiological and genetic makeup of each individual. Regular monitoring through blood work and other biomarkers is crucial to assess the effectiveness of the therapy and make necessary adjustments. This personalized and evidence-based approach ensures that peptide therapy not only supports cellular longevity but also aligns with the broader health and wellness goals of the individual.

By leveraging the power of peptides in conjunction with a healthy lifestyle and personalized medical guidance, individuals can take proactive steps towards enhancing cellular longevity and achieving a more youthful, vibrant state of health. The potential of peptides to slow the aging process and support cellular regeneration offers a promising avenue for those seeking to maintain vitality and delay the onset of age-related decline, underscoring the importance of ongoing research and clinical application in this exciting area of medicine.

8.2: KEY PEPTIDES FOR SKIN HEALTH

GHK-Cu, also known as Copper Peptide, is another powerhouse in the realm of anti-aging and skin health. Its mechanism of action is fascinating, as it involves the modulation of various cellular processes that are crucial for skin regeneration and repair. GHK-Cu has been shown to stimulate collagen and elastin production, which are the building blocks of the skin, providing it with structure and elasticity. This peptide not only enhances the skin's ability to repair itself but also has anti-inflammatory properties that can reduce the appearance of redness and irritation. The benefits of GHK-Cu extend to improving skin firmness, reducing fine lines and wrinkles, and promoting a more youthful complexion. For those looking to incorporate GHK-Cu into their skincare regimen, it is available in various forms, including serums, creams, and even injectables. The typical concentration ranges from 0.5% to 2% in topical products, which can be applied once or twice daily, depending on the product's formulation and the individual's skin tolerance. It is crucial to start with a lower concentration and gradually increase it to prevent any potential irritation. Additionally, GHK-Cu can be combined with other peptides, such as Epithalon, to create a comprehensive anti-aging protocol that addresses various aspects of skin health and cellular longevity.

Incorporating these peptides into a daily skincare routine can significantly impact skin health and appearance. However, it's important to remember that consistency is key to achieving and maintaining results. Alongside peptide use, adopting a holistic approach that includes a balanced diet rich in antioxidants, adequate hydration, sun protection, and regular exercise can further enhance skin health and contribute to overall well-being. As with any new skincare product or supplement, consulting with a healthcare professional or dermatologist before starting any new peptide protocol is advisable to ensure it aligns with individual health needs and goals.

8.2.1: Epithalon for Telomere Support

Epithalon, a synthetic tetrapeptide, has garnered attention for its potential to impact telomere length, a key factor in cellular aging and longevity. Telomeres, the protective caps at the ends of chromosomes, naturally shorten as cells divide, a process intricately linked to aging. Epithalon's role in supporting telomere extension is rooted in its ability to mimic the action of the pineal gland peptide, Epithalamin, which regulates cell replication and telomere length through the activation of telomerase, an enzyme responsible for adding DNA sequence repeats to telomeres, thereby extending their length and potentially the lifespan of cells.

The implications of Epithalon's use for anti-aging are profound, offering a novel approach to the mitigation of age-related decline at a cellular level. By promoting telomere elongation, Epithalon not only supports the maintenance of genomic stability but also enhances the replicative potential of cells, contributing to improved tissue regeneration and overall health. This mechanism of action positions Epithalon as a pivotal tool in the quest for longevity and cellular health, with studies suggesting improvements in metabolic markers, antioxidant defense systems, and neuroendocrine function following its administration.

For individuals interested in incorporating Epithalon into their anti-aging regimen, it is typically administered through subcutaneous injections, offering a direct method of delivery to the systemic circulation. The dosing of Epithalon can vary, but protocols often suggest a range from 5 to 10mg per day, administered in cycles of 10 to 20 days, followed by a period of discontinuation to prevent desensitization. It is crucial for users to source Epithalon from reputable suppliers, ensuring the purity and efficacy of the peptide to achieve desired outcomes.

The integration of Epithalon into a comprehensive anti-aging strategy should be approached with a holistic perspective, considering factors such as lifestyle, diet, and other supplements that may synergize with Epithalon's effects. For instance, antioxidants and nutrients that support DNA repair and cellular health may enhance the benefits of Epithalon, creating a multifaceted approach to longevity. Furthermore, monitoring biomarkers of aging, such as telomere length and inflammatory markers, can provide insights into the effectiveness of Epithalon and guide adjustments to the protocol.

While the promise of Epithalon in extending telomeres and improving cellular longevity is compelling, it is essential for individuals to approach its use with informed caution, recognizing the complexity of aging and the need for further research to fully understand the long-term implications of telomerase activation. As with any intervention aimed at modulating the aging process, consultation with healthcare professionals knowledgeable in peptide therapies is advisable to tailor the approach to individual health profiles and goals, ensuring safety and maximizing the potential benefits of Epithalon in the pursuit of ageless vitality.

8.2.2: GHK-Cu for Skin and Tissue Repair

GHK-Cu, also known as copper tripeptide-1, plays a pivotal role in the body's ability to repair and regenerate skin tissue, making it a cornerstone in the realm of dermatological health and anti-aging protocols. Its efficacy stems from its unique ability to upregulate genes involved in the synthesis of collagen, elastin, and glycosaminoglycans, components that are fundamental to maintaining the structural integrity and elasticity of the skin. This peptide's action extends beyond mere surface-level improvements, delving into the cellular matrix to effect changes that are both visible and enduring. The biological mechanisms triggered by GHK-Cu facilitate the removal of damaged collagen and elastin from the skin, thereby promoting the synthesis of new, healthy tissue. This process is critical in not only slowing the aging process but also in repairing the skin from various forms of damage, including UV exposure, inflammation, and certain dermatological conditions.

GHK-Cu's anti-inflammatory properties further contribute to its skin repair capabilities, reducing the cytokine-induced inflammation that can lead to accelerated aging and exacerbate skin damage. By mitigating inflammation, GHK-Cu supports the healing process, allowing for more efficient recovery from injuries and a reduction in the appearance of scars and hyperpigmentation. This makes it an invaluable tool in post-procedure care, where accelerated healing and reduced inflammation are paramount.

The versatility of GHK-Cu extends to its ability to enhance angiogenesis, the formation of new blood vessels, which is essential for delivering nutrients and oxygen to skin cells. This process is crucial for maintaining skin vitality and promoting the healing of damaged tissue. Enhanced angiogenesis, coupled with GHK-Cu's antioxidant properties, provides a dual-action approach to skin health, combating oxidative stress while supporting the skin's natural regeneration processes.

For individuals seeking to incorporate GHK-Cu into their skincare regimen, it is imperative to understand the optimal application methods and concentrations for maximum benefit. While topical formulations are the most common and accessible mode of delivery, the bioavailability of GHK-Cu can vary depending on the vehicle used. Formulations that allow for deeper penetration into the dermal layers are preferred to maximize the peptide's restorative effects. Furthermore, the combination of GHK-Cu with other peptides and active ingredients can create a synergistic effect, enhancing the overall results. Such combinations should be tailored to the individual's specific skin concerns and goals, with consideration given to the compatibility of ingredients and the desired outcome.

The application of GHK-Cu is not limited to facial skin; it can be beneficial for improving the health and appearance of skin across the body. Areas that have been damaged by sun exposure, scarring, or aging can see significant improvement with regular application of GHK-Cu-enriched products. Consistency in application is key, as the cumulative effects of GHK-Cu build over time, leading to more pronounced and lasting results. Individuals incorporating GHK-Cu into their skincare routine should also maintain a holistic approach to skin health, including adequate sun protection, hydration, and a nutrient-rich diet, to support the body's natural healing and regenerative capabilities.

While GHK-Cu offers a promising avenue for skin repair and anti-aging, it is crucial for users to source their products from reputable suppliers. The quality and purity of the peptide are paramount to achieving the desired therapeutic outcomes, underscoring the importance of selecting products that have been rigorously tested and verified for their efficacy. As the field of peptide therapy continues to evolve, GHK-Cu remains at the forefront, offering a scientifically backed solution for those seeking to enhance their skin's health and appearance through the power of peptides.

8.3: PROTOCOLS FOR ANTI-AGING AND REGENERATION

Building on the foundation of peptide protocols for anti-aging and cellular regeneration, it's crucial to understand that the efficacy of these treatments hinges on a multifaceted approach that includes lifestyle adjustments, nutritional support, and precise dosing schedules. The goal is to create an environment within the body that is conducive to the optimal functioning of these peptides, thereby enhancing their regenerative and anti-aging effects. To achieve this, a comprehensive protocol that addresses various aspects of health and well-being is necessary.

Firstly, **hydration** plays a pivotal role in ensuring that the peptides can circulate efficiently and reach their target cells. Adequate water intake supports cellular health and assists in the detoxification processes that are essential for removing cellular waste and reducing oxidative stress, a key factor in aging. Aim for at least eight glasses of water daily, more if you are active or live in a hot climate.

Chapter 8:
Anti-Aging and Cellular Regeneration

Nutrition is another cornerstone of effective peptide therapy. A diet rich in antioxidants, such as vitamins C and E, can help combat oxidative damage, while omega-3 fatty acids found in fish oil are known for their anti-inflammatory properties. Incorporating a variety of fruits, vegetables, lean proteins, and healthy fats into your diet not only supports general health but also provides the necessary building blocks for peptide synthesis and function.

Exercise should not be overlooked, as it stimulates the release of growth hormone and other beneficial hormones that can synergize with peptide therapy. A combination of resistance training and cardiovascular exercise is recommended to promote muscle synthesis, improve insulin sensitivity, and enhance cardiovascular health. Exercise also boosts mood and cognitive function, contributing to overall well-being.

When it comes to the **administration of peptides**, timing and dosage are critical. Peptides like Epithalon and GHK-Cu have specific dosing schedules that maximize their benefits. For instance, Epithalon is often administered in cycles, with a common protocol suggesting a dosage of 5 to 10mg per day for a period of 10 to 20 days, followed by a break. This cycling helps prevent desensitization and maintains the peptide's efficacy over time.

Sleep is another vital component of any anti-aging protocol. Peptides such as Epithalon have been shown to improve sleep quality, which in turn supports hormonal balance, cognitive function, and cellular repair processes that occur during deep sleep. Ensuring 7-9 hours of quality sleep per night can significantly enhance the regenerative effects of peptide therapy.

Finally, **stress management** techniques such as meditation, yoga, or even simple breathing exercises can reduce the detrimental effects of stress hormones on the body, thereby supporting the anti-aging and regenerative processes facilitated by peptide therapy. Chronic stress is a known accelerator of the aging process, making effective stress management an essential part of any anti-aging protocol.

By integrating these lifestyle and nutritional strategies with a carefully planned peptide regimen, individuals can maximize the anti-aging and regenerative potential of peptide therapy. It's important to approach this process with patience and consistency, as the benefits of peptide therapy, while profound, can take time to manifest. Regular monitoring and adjustments by a healthcare professional experienced in peptide therapy can ensure that the protocol remains effective and aligned with the individual's evolving health needs and goals.

Part 3:
Advanced Peptide Protocols

Chapter 9: Peptide Stacks for Specific Goals

9.1: Fat Loss Stacks for Optimal Weight

For individuals aiming to optimize fat loss, peptide stacks offer a targeted approach to enhancing metabolic rate and promoting the utilization of stored fat as energy. A strategic combination of peptides can synergize to achieve greater fat loss results than what might be possible with a single peptide. One such effective stack involves the use of **Tesamorelin** and **Ipamorelin**. Tesamorelin, known for its ability to reduce visceral fat, works by mimicking the natural growth hormone-releasing hormone (GHRH) which leads to an increase in the body's production of growth hormone, specifically targeting abdominal fat reduction. When combined with Ipamorelin, a selective growth hormone secretagogue, this stack not only enhances overall growth hormone levels in the body but does so without significantly increasing cortisol, a stress hormone that can lead to increased abdominal fat.

Dosage and Administration: For optimal results, Tesamorelin should be administered at a dose of 2mg per day, preferably in the evening to mimic the body's natural growth hormone release cycle. Ipamorelin, on the other hand, can be taken at a dose of 200-300mcg three times daily. This combination should be cycled for a period of 8-12 weeks to allow the body to maximize the fat-burning effects while minimizing potential desensitization to the peptides.

Another potent stack for fat loss includes the combination of **CJC-1295** and **Hexarelin**. CJC-1295 enhances growth hormone secretion, providing a steady increase in GH and IGF-1 levels without significant spikes, which can be beneficial for long-term fat loss and muscle gain. Hexarelin, one of the most potent GH-releasing peptides, further amplifies this effect. However, due to its potency, Hexarelin should be used cautiously to avoid desensitization.

Dosage and Administration: CJC-1295 is best used at a dose of 2mg once per week, combined with Hexarelin at a dose of 100mcg daily. This stack should be cycled for 8-12 weeks, followed by a 4-week break to prevent desensitization and allow the pituitary gland to recover.

It's crucial to source these peptides from reputable suppliers to ensure purity and efficacy. Additionally, while these peptide stacks can significantly aid in fat loss, they should be used in conjunction with a balanced diet and regular exercise for optimal results. Monitoring your progress through body composition analysis can help adjust dosages and cycling periods as needed to continue making progress towards your weight loss goals.

Remember, the use of peptides should always be discussed with a healthcare provider to ensure they are appropriate for your health status and goals. As with any supplement or medication, individual results can vary, and safety should always be the priority.

9.2: BRAIN FUNCTION STACKS FOR CLARITY

Chapter 9:
Peptide Stacks for Specific Goals

For individuals seeking to enhance mental clarity and focus, the strategic combination of peptides can offer significant benefits. The brain function stacks specifically designed for these purposes often include peptides known for their cognitive-enhancing properties. One such stack combines **P21** and **Cerebrolysin**, both of which have shown promise in improving cognitive functions. **P21** is a peptide that has been researched for its potential to enhance learning and memory, making it an excellent candidate for those looking to boost cognitive performance. When used in conjunction with **Cerebrolysin**, which is known for its neuroprotective and neuroregenerative properties, the synergistic effects can be particularly potent. The recommended protocol for this stack involves a careful approach to dosing, starting with lower doses of **Cerebrolysin** due to its potent effects, typically around 5ml administered intramuscularly or intravenously 2-3 times a week. **P21**, on the other hand, can be administered intranasally at a dose of 1mg per day, offering a non-invasive method of delivery that is both convenient and effective.

Another notable stack for enhancing brain function is the combination of **Semax** and **Selank**. **Semax** is primarily used for its cognitive-enhancing effects and has been shown to increase levels of brain-derived neurotrophic factor (BDNF), which plays a key role in the growth and differentiation of new neurons and synapses. **Selank**, a peptide with anxiolytic properties, complements **Semax** by reducing anxiety and stress, conditions that can significantly hinder cognitive performance. For this stack, a typical dosing strategy would involve **Semax** at 1-2mg per day administered intranasally and **Selank** at a similar dosage. This combination not only supports cognitive function but also promotes emotional stability, thereby enhancing overall mental performance.

It's crucial for users to source these peptides from reputable suppliers to ensure purity and efficacy. Additionally, individuals should consider their specific health conditions and consult with a healthcare professional before starting any new peptide regimen, particularly when aiming to address cognitive functions. Monitoring one's response to these peptides is also essential, as individual reactions can vary, and adjustments to dosages may be necessary to achieve optimal results.

While these stacks offer a promising avenue for enhancing mental clarity and focus, it's important to remember that peptides are just one part of a comprehensive approach to cognitive health. A balanced diet, regular exercise, adequate sleep, and stress management are all critical factors that work in concert with peptides to support brain function. By adopting a holistic approach that includes these lifestyle factors, individuals can maximize the cognitive benefits of peptide therapy and achieve a higher level of mental performance.

9.3: SEXUAL HEALTH PEPTIDE STACKS

In addressing the topic of **sexual health stacks for men and women**, it's essential to recognize the unique hormonal and physiological needs of each gender. For men, peptides such as **PT-141** have shown promise in enhancing libido and sexual function. PT-141 works by activating the melanocortin receptors in the brain, which are involved in sexual arousal. A typical protocol might involve a subcutaneous injection of 1-2mg of PT-141 approximately 45 minutes before anticipated sexual activity. However, it's crucial to start with the lower end of the dosage range to assess tolerance and gradually adjust as needed.

For women, the peptide **Bremelanotide**, also known as PT-141, can be used to improve sexual desire and arousal disorders. Unlike its application in men, women may find a lower dose effective, starting at 0.75mg to 1mg. The method of administration and timing similar to that recommended for men applies here, with adjustments made based on individual response and tolerance.

Both genders can benefit from the addition of **Oxytocin**, a peptide known for its role in social bonding, sexual reproduction, and during and after childbirth. Oxytocin can enhance emotional connection and physical intimacy between partners. A nasal spray form, dosed at 10-20 IU, can be used shortly before intimate activities to foster a deeper emotional and physical bond.

It's also worth noting the importance of **GHRP-6** and **CJC-1295** for both men and women, not only for their growth hormone-releasing properties but also for their potential to improve energy, vitality, and overall well-being, which indirectly supports a healthy sex life. A protocol involving GHRP-6 might include 100-300 mcg taken subcutaneously 2-3 times daily, while CJC-1295 with DAC could be administered at 2mg once weekly to optimize growth hormone levels and support sexual health.

The synergy between these peptides can offer a comprehensive approach to enhancing sexual health, but it's imperative to source these compounds from reputable suppliers to ensure purity and efficacy. Furthermore, individuals should consult with a healthcare professional to tailor these protocols to their specific health status, goals, and any potential contraindications. Monitoring one's response to these peptides is crucial, as individual reactions can vary, necessitating adjustments to dosages or protocols to achieve the desired outcomes.

Incorporating these peptides into a broader lifestyle approach that includes a balanced diet, regular exercise, and stress management techniques can further enhance sexual health and overall well-being. While peptides can offer significant benefits, they are

Chapter 9:
Peptide Stacks for Specific Goals

most effective when part of a holistic health strategy that addresses all aspects of an individual's life.

9.4: SKIN AND COSMETIC PEPTIDE STACKS

For individuals aiming to enhance their skin's appearance and maintain a youthful glow, the strategic combination of certain peptides can offer transformative results. The key to leveraging the power of peptides in skincare lies in understanding which peptides work best for specific skin concerns and how they can be combined for maximum effect. **Matrixyl (Palmitoyl Pentapeptide-4)**, for instance, is renowned for its ability to stimulate collagen production, thereby reducing the appearance of fine lines and wrinkles. When paired with **Argireline (Acetyl Hexapeptide-8)**, which targets expression wrinkles by inhibiting muscle movement, the duo provides a comprehensive anti-aging solution. A recommended protocol might involve applying a serum containing both peptides in the morning and evening, following a thorough cleansing routine.

Another potent peptide for skin health is **Copper Peptide (GHK-Cu)**, which not only promotes collagen and elastin production but also enhances the skin's ability to repair itself. For those dealing with skin damage or signs of aging, incorporating GHK-Cu into their skincare regimen can lead to visible improvements in skin texture and firmness. Combining GHK-Cu with **Hyaluronic Acid** in a serum form ensures deep hydration while maximizing the regenerative effects of the peptides on the skin.

For individuals concerned with hyperpigmentation and uneven skin tone, **Alpha-Arbutin** and **Peptide-38** can be effective. Alpha-Arbutin is a powerful brightening agent that works to fade dark spots and inhibit melanin production, while Peptide-38 stimulates the skin's renewal process, smoothing out texture and tone. A blend of these ingredients can be applied as a spot treatment or all over the face to achieve a more even complexion.

It's critical to source these peptides from reputable suppliers to ensure their purity and efficacy. Additionally, users should pay close attention to the concentration of peptides in their chosen products, as this can significantly impact their effectiveness. Starting with lower concentrations and gradually increasing, based on skin tolerance, can help minimize potential irritation while allowing the skin to adapt to the powerful effects of peptides.

Moreover, while peptides can significantly improve skin health and appearance, they should be part of a broader skincare and health regimen. Protecting the skin from sun damage with a high-SPF sunscreen, maintaining a balanced diet rich in antioxidants, and ensuring adequate hydration are all essential steps that complement the use of peptide stacks. Regular exercise and sufficient sleep also play a crucial role in skin health, supporting the body's natural repair processes and enhancing the overall benefits of peptide therapy.

Incorporating these peptides into a consistent skincare routine, along with a healthy lifestyle, can lead to significant improvements in skin appearance and texture. Patience and persistence are key, as the benefits of peptides accumulate over time, leading to a more radiant, youthful complexion.

Chapter 10: Peptides, Supplements, and Lifestyle

10.1: Nutritional Support for Peptide Efficacy

Optimizing the efficacy of peptide protocols through nutritional support is a critical aspect of maximizing the benefits of peptide therapy. A well-structured diet can significantly enhance the body's response to peptides, promoting better health outcomes. Essential nutrients play a pivotal role in this process, acting as cofactors that support the biochemical pathways activated by peptides. For instance, **vitamin C** is crucial for collagen synthesis, enhancing the effectiveness of peptides aimed at skin health and anti-aging. Incorporating foods rich in vitamin C, such as oranges, strawberries, bell peppers, and kale, can provide the necessary support for these peptides to perform optimally.

Similarly, **zinc** is another nutrient that supports the immune system and wound healing, making it a valuable addition to the diet when using peptides like Thymosin Alpha-1 for immune modulation. Foods high in zinc, including beef, pumpkin seeds, lentils, and chickpeas, can bolster the immune-enhancing effects of certain peptides. Moreover, **omega-3 fatty acids**, found in fish like salmon and mackerel, as well as in flaxseeds and walnuts, are known for their anti-inflammatory properties. These fatty acids can complement peptides aimed at reducing inflammation, supporting muscle recovery, and improving cognitive function.

Magnesium is another nutrient that supports muscle function and recovery, sleep quality, and nervous system health. Including magnesium-rich foods such as spinach, almonds, and black beans in your diet can enhance the benefits of peptides designed for muscle recovery and sleep enhancement. Additionally, **B vitamins**, particularly B12 and B6, are vital for energy metabolism and cognitive function. Foods like chicken, eggs, and avocados are rich in B vitamins, supporting the efficacy of peptides that target energy levels and mental clarity.

To support the overall effectiveness of peptide protocols, it's also essential to maintain adequate hydration. Water plays a crucial role in transporting nutrients, removing waste, and facilitating cellular processes. Ensuring a sufficient intake of water can enhance the body's ability to utilize peptides effectively.

Incorporating a diverse range of fruits, vegetables, lean proteins, healthy fats, and whole grains into your diet provides a broad spectrum of nutrients that can support the various functions of peptides. Tailoring your diet to include these key nutrients can create a synergistic effect, amplifying the benefits of peptide therapy. It's important to remember that while peptides can offer significant health advantages, they are most effective when used as part of a comprehensive approach to health that includes proper nutrition, regular exercise, and adequate sleep. By focusing on a balanced diet that supports the specific actions of peptides, individuals can achieve optimal results, enhancing vitality, recovery, sleep quality, and cognitive function.

10.2: SUPPLEMENT PAIRINGS FOR PEPTIDE RESULTS

Chapter 10:
Peptides, Supplements, and Lifestyle

To further enhance the outcomes of peptide protocols, integrating specific supplements can play a pivotal role in maximizing benefits and addressing the holistic needs of the body. **Coenzyme Q10 (CoQ10)**, a powerful antioxidant, supports cellular energy production and can be particularly beneficial when using peptides aimed at increasing energy levels and improving metabolic functions. A daily dosage of 100-200 mg of CoQ10 is recommended to complement energy-boosting peptides, aiding in the optimization of mitochondrial function and overall vitality.

Omega-3 fatty acids are another crucial supplement, known for their anti-inflammatory properties and support of cognitive health. When paired with cognitive-enhancing peptides, omega-3s, particularly EPA and DHA, help in maintaining fluidity of cell membranes and promoting brain health. A daily intake of 1000-2000 mg of EPA and DHA combined can significantly enhance the cognitive benefits of peptides like Semax and Selank, supporting mental clarity and focus.

Vitamin D is essential for immune function, bone health, and mood regulation. It works synergistically with peptides that modulate the immune system, such as Thymosin Alpha-1. For individuals using peptides for immune support, supplementing with 2000-5000 IU of vitamin D daily can help in optimizing the immune-boosting effects, especially in regions with limited sunlight exposure.

Magnesium plays a critical role in muscle function, nervous system regulation, and sleep quality. It complements peptides aimed at improving sleep and muscle recovery, such as those used in post-workout protocols. A dosage of 200-400 mg of magnesium before bedtime can enhance the sleep-inducing effects of peptides, promoting deeper and more restorative sleep.

Probiotics are beneficial for gut health, which is intricately linked to overall wellness, immune function, and even mental health. Incorporating a high-quality probiotic supplement can amplify the benefits of peptides used for immune support and cognitive function by maintaining a healthy gut microbiome. Look for probiotics with a broad spectrum of strains and a colony-forming unit (CFU) count in the billions to ensure efficacy.

Creatine is well-known for its benefits in increasing strength, muscle mass, and performance in resistance training. When combined with muscle-building peptides, such as those targeting growth hormone release, creatine can further enhance muscle synthesis and recovery. A standard dosage of 5g per day is recommended for individuals looking to maximize their gains from peptide protocols focused on muscle growth.

Curcumin, the active compound in turmeric, offers potent anti-inflammatory and antioxidant benefits. It can be particularly useful when used alongside peptides that reduce inflammation or are involved in recovery protocols. Supplementing with 500-1000 mg of curcumin, preferably with piperine to enhance absorption, can help in reducing inflammation and supporting the healing process.

Incorporating these supplements into your peptide regimen can significantly enhance the effectiveness of the peptides, providing a more comprehensive approach to health optimization. It's important to source high-quality supplements and consult with a healthcare professional before starting any new supplement regimen, especially to ensure compatibility with your specific health conditions and goals. By carefully selecting supplements that complement your peptide protocols, you can achieve a synergistic effect that supports your overall health and well-being, maximizing the benefits of your peptide therapy.

10.3: LIFESTYLE CHANGES FOR PEPTIDE BENEFITS

Adopting lifestyle modifications is crucial for maximizing the benefits of peptide therapy, focusing on enhancing sleep quality and managing stress effectively. These changes not only support the physiological effects of peptides but also contribute to overall well-being and health optimization. **Sleep** is foundational to health, serving as a critical period for recovery, hormonal regulation, and cognitive function. To improve sleep quality, establish a consistent sleep schedule by going to bed and waking up at the same times every day, even on weekends. This regularity reinforces your body's sleep-wake cycle, promoting deeper, more restorative sleep. Create a bedtime routine that signals to your body it's time to wind down; this could include activities such as reading, taking a warm bath, or practicing relaxation exercises. Limit exposure to screens and blue light at least an hour before bedtime, as they can disrupt melatonin production and interfere with the ability to fall asleep. Ensure your sleeping environment is conducive to rest, keeping the room dark, quiet, and at a comfortable temperature. Consider using blackout curtains, white noise machines, and high-quality mattresses and pillows to enhance your sleep environment.

Stress management is equally important in maximizing peptide benefits, as chronic stress can undermine health and negate the positive effects of peptides. Incorporate regular physical activity into your routine, aiming for at least 30 minutes of moderate exercise most days of the week. Exercise not only improves physical health but also reduces stress and enhances mood by releasing endorphins. Practice mindfulness and relaxation techniques such as meditation, deep breathing exercises, or yoga. These practices can help lower stress levels, reduce blood pressure, and improve overall emotional well-being. Develop a support network of friends, family, and possibly professionals who can provide emotional support and practical advice when needed. Learning to say no and setting healthy boundaries can also significantly reduce stress levels, preventing overcommitment and burnout.

Dietary habits play a supportive role in enhancing the efficacy of peptides. Focus on a balanced diet rich in whole foods, incorporating a variety of fruits, vegetables, lean proteins, and healthy fats. This nutritional foundation provides the necessary vitamins, minerals, and antioxidants that support the body's natural processes, including those influenced by peptide therapy. Limit intake of processed foods, sugars, and excessive caffeine, as these can contribute to inflammation, disrupt sleep patterns, and exacerbate stress.

Hydration is another key factor; ensure adequate water intake to support cellular health, detoxification, and overall bodily functions. Proper hydration is essential for maintaining the balance of bodily fluids, facilitating nutrient delivery to cells, and supporting kidney function.

Mindset and emotional well-being are integral to the success of any health optimization strategy, including peptide therapy. Cultivate a positive outlook and practice gratitude, focusing on the aspects of your life that bring joy and fulfillment. Engaging in hobbies and activities that you love can significantly improve your quality of life and reduce stress.

Finally, **regular monitoring and adjustment** of your lifestyle habits and peptide protocols are necessary to ensure you are achieving the desired outcomes. Keep a journal or use apps to track your sleep, stress levels, diet, and exercise routines, making adjustments as needed based on your observations and how you feel. This proactive approach allows you to fine-tune your lifestyle modifications and peptide use, ensuring they work synergistically to achieve your health and wellness goals.

By integrating these lifestyle modifications, you can create a supportive environment that maximizes the benefits of peptide therapy, leading to improved vitality, recovery, sleep quality, and mental clarity. Remember, consistency is key, and making these changes a permanent part of your lifestyle will provide the best foundation for health optimization and the effective use of peptides.

10.3.1: Diet and Exercise Adjustments

Optimizing peptide absorption and efficacy goes beyond the peptides themselves; it involves a holistic approach that includes diet and exercise adjustments. A well-structured diet can significantly enhance the body's response to peptide therapy. For instance, incorporating foods rich in protein and amino acids can provide the necessary building blocks for peptides to work effectively. Foods such as lean meats, fish, eggs, and legumes are excellent sources. Additionally, including a variety of fruits and vegetables ensures an ample supply of vitamins and minerals, supporting overall health and optimizing the body's environment for peptide function.

Hydration plays a crucial role in peptide efficacy. Adequate water intake is essential for maintaining cellular health and facilitating the transport of peptides through the bloodstream. It's recommended to drink at least eight 8-ounce glasses of water daily, although this may need to be adjusted based on individual activity levels and environmental conditions.

Regarding exercise, engaging in regular physical activity can enhance the body's sensitivity to peptides. Exercise stimulates various physiological processes that can synergize with peptide therapy, such as improving blood circulation, which allows peptides to be more efficiently distributed throughout the body. Resistance training, in particular, can be beneficial as it helps in building muscle mass, a common goal for those using peptides for muscle growth and recovery. However, it's important to balance exercise types and intensity according to one's health status and goals. Overtraining can lead to increased stress levels and inflammation, potentially counteracting the benefits of peptide therapy.

For those focusing on peptides for weight management, incorporating a mix of cardiovascular exercises can aid in fat loss by increasing calorie burn. Meanwhile, flexibility and mobility exercises, such as yoga or Pilates, can support recovery and reduce the risk of injury, ensuring that peptide protocols aimed at muscle recovery and growth are more effective.

It's also worth noting that the timing of nutrient intake and exercise can influence peptide efficacy. Consuming protein-rich foods or amino acid supplements shortly after a workout can provide the necessary materials for muscle repair and growth, enhancing the regenerative effects of certain peptides. Similarly, engaging in exercise during times when peptide levels are naturally higher, such as in the morning for some individuals, may amplify the benefits of both exercise and peptide therapy.

Ultimately, the integration of diet and exercise adjustments into a peptide protocol requires a personalized approach. Factors such as age, gender, health status, and specific goals should all be considered when designing a regimen that includes dietary

Chapter 10:
Peptides, Supplements, and Lifestyle

changes and exercise. Consulting with a healthcare provider or a nutrition and fitness expert can provide valuable insights and help tailor a plan that maximizes the benefits of peptides while ensuring safety and overall health.

10.3.2: Stress Management for Peptide Performance

Managing stress is essential for optimizing the performance of peptide protocols, as chronic stress can significantly hinder the body's ability to respond to peptides effectively. Stress triggers the release of cortisol, a hormone that can disrupt hormone balance, impair immune function, and reduce the efficacy of peptides aimed at improving health and vitality. Therefore, incorporating stress management techniques into your lifestyle is not just beneficial but necessary for maximizing the benefits of peptide therapy.

Mindfulness and meditation have been shown to be effective tools for reducing stress. These practices help in calming the mind, reducing cortisol levels, and enhancing focus, thereby creating a more conducive environment for peptides to work. Regular mindfulness exercises can improve mental clarity and reduce the feelings of overwhelm, making it easier to maintain a consistent peptide protocol. Meditation, even if practiced for as little as ten minutes a day, can significantly lower stress levels and improve overall well-being.

Another critical aspect of stress management is sleep. Quality sleep is paramount for the body's recovery and regeneration processes, which peptides often aim to support. Establishing a regular sleep schedule, optimizing your sleep environment for comfort and darkness, and avoiding stimulants such as caffeine late in the day can enhance sleep quality. Additionally, certain peptide protocols that support deep sleep can be more effective when combined with good sleep hygiene, thus creating a synergistic effect that further reduces stress and enhances recovery and vitality.

Physical activity is also a potent stress reducer. Regular exercise, particularly aerobic exercises like walking, running, or cycling, can significantly lower stress levels by releasing endorphins, the body's natural painkillers and mood elevators. Exercise also helps in improving sleep quality and boosting confidence, which in turn, can make it easier to handle stress. It's important to find a form of exercise you enjoy and can perform regularly without causing undue strain or injury, as this ensures sustainability and consistency in your stress management routine.

Nutrition plays a role in managing stress as well. Consuming a balanced diet rich in antioxidants, vitamins, and minerals can help in combating oxidative stress and supporting the body's natural stress response. Foods high in magnesium, for example, such as leafy greens, nuts, and seeds, can help in relaxing muscles and reducing tension. Omega-3 fatty acids, found in fish and flaxseeds, have been shown to reduce the levels of cortisol and adrenaline during stressful situations. Incorporating these nutrients into your diet can support your body's resilience to stress, enhancing the effectiveness of peptide protocols.

Finally, building a support network of friends, family, and professionals can provide emotional support and practical advice for managing stress. Sharing your experiences and challenges with others who understand can be incredibly relieving and provide a sense of belonging and community. For those who find stress overwhelming, seeking professional help from a therapist or counselor can offer strategies to manage stress effectively, tailoring approaches to individual needs and situations.

Incorporating these stress management techniques into your daily routine can significantly enhance the effectiveness of peptide protocols. By reducing stress, you create an optimal internal environment that allows peptides to work more efficiently, thereby maximizing health benefits and achieving your wellness goals.

Chapter 11:
Monitoring Your Progress

11.1: BIOMARKERS AND HEALTH METRICS

Monitoring the effectiveness of peptide protocols is crucial for optimizing health outcomes. Biomarkers and health metrics serve as tangible indicators of physiological changes and can guide adjustments to peptide regimens for improved results. **Inflammatory markers**, such as C-reactive protein (CRP), provide insight into the body's inflammatory response, which can be modulated by certain peptides to enhance recovery and immune function. A significant decrease in CRP levels may indicate effective immune modulation and a reduction in systemic inflammation. **Hormone levels**, including insulin-like growth factor 1 (IGF-1) and growth hormone (GH), are directly influenced by peptides like CJC-1295 and Ipamorelin. Elevations in these hormones can signal enhanced anabolic activity, contributing to muscle growth and fat loss. Monitoring these levels through blood tests can help assess the efficacy of peptide protocols aimed at boosting GH and IGF-1.

Blood glucose and insulin sensitivity metrics are essential for individuals using peptides for metabolic health and weight management. Improved insulin sensitivity and stable blood glucose levels can indicate successful metabolic regulation by peptides such as AOD-9604. Regular monitoring through fasting blood glucose tests and HbA1c levels can provide feedback on metabolic health improvements. **Sleep quality**

improvements, while subjective, can be tracked using wearable devices that monitor sleep stages and duration. Peptides like DSIP (Delta sleep-inducing peptide) aim to enhance deep sleep, and an increase in deep sleep duration observed through sleep tracking devices can suggest effective sleep optimization.

Muscle mass and body composition changes can be quantified using DEXA scans or bioelectrical impedance analysis (BIA). These methods offer precise measurements of body fat percentage, lean muscle mass, and bone density, providing objective data on the physical changes induced by peptides promoting muscle growth and fat loss. **Cognitive function** improvements, though challenging to quantify, can be assessed through standardized cognitive tests and mood questionnaires. Peptides like Semax and Selank, known for their nootropic effects, may lead to measurable enhancements in memory, focus, and mood stability.

Skin health can be evaluated through dermatological assessments and photographic documentation over time. Peptides such as GHK-Cu, which support skin regeneration and collagen production, may result in visible improvements in skin texture, elasticity, and overall appearance. Documenting these changes provides concrete evidence of anti-aging effects. **Immune function** can be monitored through white blood cell counts and immunoglobulin levels, offering insights into the body's defense mechanisms. Peptides like Thymosin Alpha-1, known for their immune-boosting properties, can lead to optimized immune responses, as reflected in these metrics.

In conclusion, tracking biomarkers and health metrics is indispensable for personalizing peptide protocols and maximizing their benefits. Regular monitoring, in conjunction with healthcare provider consultations, ensures that peptide use is safe, effective, and aligned with individual health goals.

11.2: TOOLS AND APPS FOR TRACKING PEPTIDES

In the realm of peptide therapy, leveraging technology for tracking and monitoring results is not just advantageous; it's essential for tailoring protocols to individual needs and maximizing outcomes. A variety of tools and apps have emerged, designed specifically to assist users in logging their peptide usage, physical changes, and biomarker data, thereby providing a comprehensive overview of their progress and the effectiveness of their peptide protocols. **MyFitnessPal** is a widely recognized app that, while primarily focused on diet and exercise, offers robust features for tracking nutrient intake and physical activity levels. This can be particularly useful for individuals using peptides for weight management or muscle building, as it allows for the monitoring of protein intake and exercise, which are critical components of these goals. **Fitbit** and **Garmin Connect** are notable for their sleep tracking capabilities. Given the importance of sleep in recovery and overall health, users engaging in peptide protocols aimed at enhancing sleep quality can benefit from the detailed analysis these platforms offer regarding sleep stages and duration. This data can be invaluable in assessing the impact of peptides like DSIP on sleep patterns.

Strava offers an engaging platform for those focused on cardiovascular health and performance, providing detailed metrics on physical activities such as running, cycling, and swimming. For individuals utilizing peptides to enhance athletic performance, Strava can serve as a motivational tool, enabling users to set goals, track progress, and even compare their performance with that of others. **Cronometer** is another comprehensive tool that excels in the detailed tracking of micronutrients and biometrics. Users can input daily food intake and monitor a wide array of nutrients, which is particularly beneficial for ensuring that dietary intake supports and enhances the effects of peptide protocols. Additionally, Cronometer allows for the logging of biometric data, such as blood pressure and blood glucose levels, offering a holistic view of one's health status.

Headspace and **Calm** are leading apps in the domain of mental wellness, providing resources for meditation, stress management, and sleep. For individuals incorporating peptides into their regimen for cognitive enhancement or stress reduction, these apps can complement peptide therapy by promoting mental clarity, reducing stress levels, and improving sleep quality. **LabCorp** and **Quest Diagnostics** have developed apps that facilitate the easy scheduling of lab tests and access to test results. For those monitoring biomarkers to gauge the efficacy of peptide protocols, these apps streamline the process of obtaining and interpreting lab data, making it easier to adjust protocols based on objective health metrics.

Incorporating these tools and apps into a peptide protocol not only aids in tracking progress but also empowers individuals to make data-driven decisions about their health. By meticulously logging peptide usage, dietary habits, exercise routines, sleep patterns, and biomarker data, users can identify patterns and correlations that inform the optimization of their peptide protocols. This approach ensures a personalized strategy that aligns with individual goals, health status, and lifestyle, ultimately enhancing the potential for achieving desired outcomes with peptide therapy.

Part 4:
Personalized Future Approaches

Chapter 12: Personalized Peptide Protocols

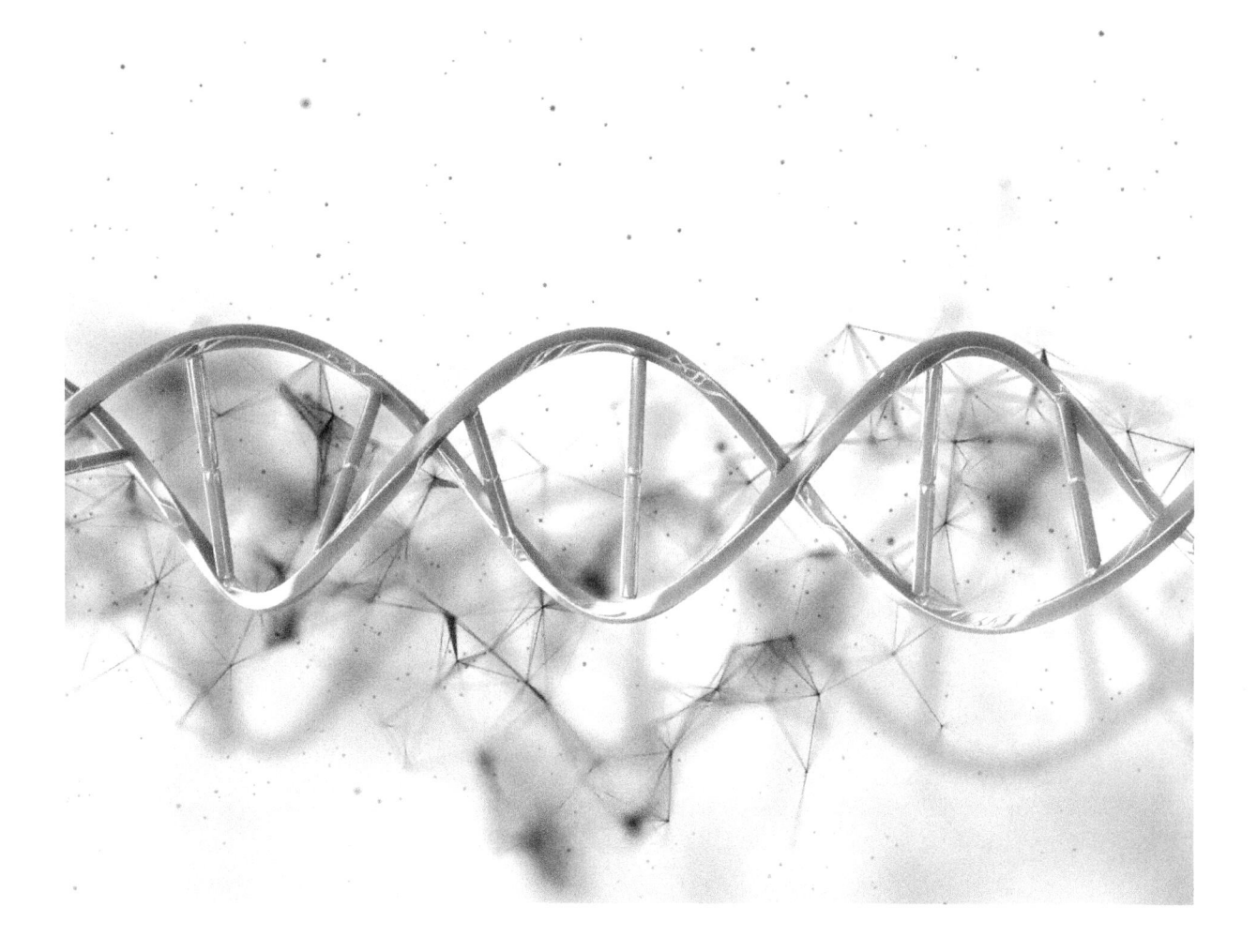

12.1: Genetic and Personalized Peptide Use

Utilizing DNA testing to tailor peptide protocols offers a groundbreaking approach to health optimization, allowing for a level of personalization previously unattainable. By analyzing an individual's genetic makeup, healthcare providers can identify specific genetic markers that influence how one's body might respond to various peptides. This genetic insight enables the creation of highly customized peptide regimens that align with one's unique biological predispositions, enhancing efficacy while minimizing potential adverse reactions. For instance, certain genetic variations can affect the body's ability to synthesize and metabolize peptides, impacting their overall effectiveness and safety. By identifying these variations, a healthcare provider can adjust peptide types, dosages, and administration schedules to better suit the individual's genetic profile.

Moreover, genetic testing can reveal predispositions to certain health conditions or deficiencies that peptides can address. For example, if an individual's genetic analysis indicates a predisposition towards lower bone density, peptides known to support bone health, such as those mimicking the action of growth hormone, could be prioritized in their treatment plan. Similarly, genetic markers related to inflammation and immune response can guide the inclusion of peptides that modulate immune function or target inflammatory processes, offering targeted intervention for autoimmune conditions or chronic inflammation.

Biomarker monitoring complements genetic analysis by providing ongoing, tangible feedback on the body's response to peptide therapy. Regular assessment of biomarkers related to the targeted health outcomes allows for real-time adjustments to peptide protocols. This dynamic approach ensures that the regimen remains aligned with the individual's evolving health status and goals. Key biomarkers vary depending on the objectives of peptide use but can include hormone levels for those targeting hormonal balance, inflammatory markers for individuals focusing on immune modulation or inflammation reduction, and markers of bone and muscle health for protocols aimed at enhancing physical performance and recovery.

Incorporating genetic and biomarker data into peptide protocol design not only maximizes the potential benefits but also underscores the importance of a personalized approach to health and wellness. It represents a shift from a one-size-fits-all methodology to a more nuanced, individualized strategy that considers the complex interplay between genetics, lifestyle, and health. As such, individuals interested in exploring peptide therapy are encouraged to seek out healthcare providers with expertise in genetic analysis and personalized medicine. These professionals can guide the selection of appropriate genetic tests, interpret the results, and develop a peptide protocol that is truly tailored to the individual's genetic blueprint and health objectives.

The integration of genetic testing and biomarker monitoring into peptide therapy exemplifies the cutting-edge of personalized medicine. It offers individuals an unprecedented opportunity to harness their genetic information for targeted health optimization, embodying a proactive and informed approach to wellness. As research in this area continues to evolve, it is anticipated that the use of genetic and biomarker data to inform peptide therapy will become increasingly sophisticated, further enhancing the efficacy and safety of these powerful molecules in promoting health and vitality.

12.1.1: Utilizing DNA for Personalized Peptide Protocols

Utilizing DNA testing to tailor peptide protocols represents a cutting-edge approach in the realm of personalized medicine, offering a pathway to significantly enhance the efficacy of peptide therapies. This method hinges on the understanding that genetic variations among individuals can influence how one responds to different peptides, affecting both the efficacy and safety of these compounds. By analyzing an individual's genetic makeup, healthcare professionals can identify specific genetic markers that may predict responses to certain peptides, thereby enabling the customization of peptide protocols to align with one's unique genetic profile.

The process begins with a comprehensive DNA analysis, focusing on genes related to peptide metabolism, receptor sensitivity, and potential predispositions to side effects. This genetic screening can uncover insights into how efficiently an individual's body might metabolize various peptides, which receptors might be more responsive, and what dosage ranges could be optimal or harmful. For instance, variations in the GH1 gene might influence how effectively a person responds to growth hormone-releasing peptides, guiding the selection and dosing of peptides like GHRP-6 or CJC-1295 for muscle growth or fat loss.

Following the identification of relevant genetic markers, a personalized peptide protocol can be developed. This tailored approach not only aims to optimize the therapeutic outcomes—such as enhanced muscle recovery, improved cognitive function, or accelerated fat loss—but also minimizes the risk of adverse reactions. For example, if genetic testing reveals a higher sensitivity to certain peptides that could potentially elevate cortisol levels, the protocol can be adjusted to include peptides that counteract this effect or to modify the dosing schedule to mitigate unwanted side effects.

Moreover, the integration of DNA testing into peptide protocol development underscores the importance of ongoing monitoring and adjustment. As our understanding of genetics and peptide science evolves, so too should the personalized protocols. Regular follow-ups and possibly re-testing can ensure that the peptide regimen remains aligned with the individual's changing health status, lifestyle factors, and long-term goals.

Incorporating DNA testing into the customization of peptide protocols represents a profound shift towards truly personalized health optimization strategies. It leverages the power of genetic insights to tailor interventions that are not only more effective but also safer for the individual. This approach exemplifies the potential of personalized medicine to transform how we enhance human health and performance, moving beyond one-size-fits-all solutions to embrace the complexity and uniqueness of each person's genetic blueprint.

12.1.2: Biomarker Monitoring for Optimized Outcomes

Biomarker monitoring serves as a critical component in the personalized peptide protocol, offering a dynamic approach to optimizing therapeutic outcomes. This method involves the regular measurement of specific biomarkers in the body, which can provide invaluable insights into an individual's response to peptide therapy. Biomarkers, essentially, are biological molecules found in blood, other body fluids, or tissues that are a sign of a normal or abnormal process, or of a condition or disease. They can be used to see how well the body responds to a treatment for a disease or condition.

For individuals embarking on peptide protocols, understanding which biomarkers to monitor and how to interpret the changes can significantly enhance the efficacy of their regimen. For instance, monitoring inflammation markers such as C-reactive protein (CRP) can help gauge the anti-inflammatory effects of certain peptides, adjusting dosages or combinations as needed to achieve optimal results. Similarly, tracking hormone levels, including insulin-like growth factor 1 (IGF-1), can provide insights into the effectiveness of peptides designed to stimulate growth hormone production, offering a basis for fine-tuning the therapy to better suit the individual's physiological needs.

Moreover, the monitoring of biomarkers related to cognitive function, such as brain-derived neurotrophic factor (BDNF), can aid in assessing the impact of peptides on brain health and cognitive performance. This allows for the customization of peptide protocols aimed at enhancing cognitive clarity and mental focus, ensuring that the selected peptides are yielding the desired outcomes.

The process of biomarker monitoring typically involves periodic testing, which can be conducted through blood tests, urine tests, or other relevant medical examinations. The frequency and type of tests will depend on the specific peptides being used, the goals of the therapy, and the individual's unique health profile. It is crucial for these tests to be interpreted by healthcare professionals who can provide expert guidance on adjusting the peptide protocol based on the results. This collaborative approach ensures that the therapy remains aligned with the individual's evolving health status and therapeutic goals.

To effectively implement biomarker monitoring in personalized peptide protocols, individuals should work closely with their healthcare providers to identify the most relevant biomarkers for their specific health objectives. This may involve a comprehensive initial assessment to establish baseline levels for selected biomarkers, followed by regular monitoring to track changes over time. The insights gained from this monitoring can then be used to make informed decisions about adjusting dosages, cycling protocols, or incorporating additional peptides or supplements to enhance the therapy's effectiveness.

The integration of biomarker monitoring into personalized peptide protocols represents a powerful strategy for maximizing the benefits of peptide therapy. By providing a mechanism for ongoing assessment and adjustment, this approach enables individuals to achieve more targeted and effective outcomes, tailored to their unique physiological makeup and health objectives. As the field of peptide science continues to evolve, the role of biomarker monitoring in personalized medicine is poised to become increasingly central, offering a pathway to more precise, effective, and individualized therapeutic interventions.

12.2: ADAPTING PROTOCOLS TO LIFESTYLE AND GOALS

Adapting peptide protocols to align with individual lifestyles and goals requires a nuanced understanding of how various factors such as diet, activity level, stress, and sleep patterns can influence peptide efficacy. The first step in this process involves a thorough assessment of one's daily routine, identifying areas where adjustments can be made to support the desired outcomes of peptide therapy. For instance, individuals aiming for muscle growth or recovery might need to increase their protein intake and ensure they are getting enough sleep to facilitate tissue repair. Similarly, those seeking cognitive enhancements from peptides should consider optimizing their diet to include nutrients that support brain health, such as omega-3 fatty acids, while also implementing strategies to manage stress, as chronic stress can undermine cognitive function.

Activity level plays a critical role in how the body responds to peptides. Regular exercise, particularly resistance training, can enhance the body's sensitivity to peptides involved in muscle synthesis, such as GHRP-6 or CJC-1295. However, it's important to balance exercise intensity and recovery, as overtraining can lead to elevated cortisol levels, which may counteract some of the beneficial effects of peptides. Tailoring exercise routines to include a mix of high-intensity and restorative practices, like yoga or meditation, can help maintain an optimal balance for peptide efficacy.

Dietary habits also significantly impact the success of peptide protocols. Nutrients from food can either support or hinder the body's response to peptides. For example, diets high in refined sugars and processed foods can increase inflammation, potentially diminishing the effectiveness of peptides designed to promote tissue repair or cognitive function. Conversely, a diet rich in whole foods, antioxidants, and anti-inflammatory compounds can enhance the body's receptivity to peptides. Incorporating foods that support the production of growth hormone, such as those high in arginine (nuts, seeds, and legumes) and glutamine (beef, chicken, fish, dairy products), can be particularly beneficial for individuals using peptides for muscle growth or anti-aging purposes.

Stress management is another crucial element in optimizing peptide protocols. Chronic stress elevates cortisol levels, which can interfere with the desired effects of peptides, especially those targeting the immune system, muscle recovery, and cognitive function. Techniques such as mindfulness meditation, deep-breathing exercises, and regular physical activity can help manage stress levels, thereby enhancing the overall effectiveness of peptide therapy.

Sleep quality cannot be overstated in its importance for maximizing peptide benefits. Poor sleep can disrupt hormone balance, impair recovery processes, and reduce cognitive function, undermining the goals of peptide therapy. Ensuring adequate, high-quality sleep supports the body's natural growth hormone cycles, which is essential for the efficacy of many peptide protocols. Strategies to improve sleep include establishing a regular sleep schedule, reducing exposure to blue light before bedtime, and creating a restful sleeping environment.

In summary, adapting peptide protocols to individual lifestyles and goals involves a holistic approach that encompasses diet, exercise, stress management, and sleep optimization. By tailoring these lifestyle factors to support specific peptide therapies, individuals can significantly enhance the effectiveness of their protocols, achieving more pronounced and sustainable results. Collaboration with healthcare professionals can provide additional insights and adjustments, ensuring that peptide use is both safe and optimally aligned with personal health objectives.

Chapter 13:
Future Trends in Peptide Therapy

13.1: CUTTING-EDGE RESEARCH AND PEPTIDES

The realm of **cutting-edge research** in peptide therapy is rapidly evolving, pushing the boundaries of traditional medicine and opening new avenues for treating complex diseases. Among the most promising developments are **experimental peptides** that target a range of conditions, from neurodegenerative disorders to autoimmune diseases, showcasing the potential of peptides to revolutionize healthcare. One such peptide,

BPC-157, known for its remarkable healing properties, has shown efficacy in accelerating wound healing and tissue regeneration. Its mechanism involves promoting the formation of new blood vessels, a process known as angiogenesis, and facilitating the repair of damaged tissues, making it a subject of intense study for its applications in recovery from injuries and surgeries.

Another area of significant interest is the development of **peptides for neuroprotection**, such as **Dihexa**, which has been found to have potent neurogenic effects. Dihexa is a peptide that can cross the blood-brain barrier, a critical factor in treating brain-related conditions, and it binds to hepatocyte growth factor to induce synaptic plasticity, crucial for learning and memory. This peptide has shown promise in animal models for conditions like Alzheimer's disease, suggesting a potential pathway to treat cognitive decline.

The **immunomodulatory peptides** are another frontier in peptide research, with peptides like **Thymosin Beta-4 (TB-500)** drawing attention for their ability to modulate the immune system. TB-500 has been shown to reduce inflammation and promote cell migration, properties that can be leveraged to treat autoimmune diseases and improve overall immune resilience. Its ability to regulate the actin cytoskeleton of cells also highlights its potential in wound repair and tissue regeneration.

In the fight against cancer, **oncolytic peptides** represent a novel approach that selectively targets cancer cells without harming normal tissue. These peptides can induce apoptosis in cancer cells, disrupt their metabolism, or inhibit angiogenesis, effectively starving the tumor of its blood supply. The specificity and low toxicity profile of oncolytic peptides make them an attractive alternative to traditional chemotherapy and radiation treatments, with several candidates currently undergoing clinical trials.

The integration of **peptide therapy with gene editing technologies** like CRISPR-Cas9 offers another exciting avenue for research. By delivering gene-editing tools directly into cells using peptides, scientists can correct genetic mutations at their source, offering hope for hereditary diseases that were once thought to be untreatable. This approach combines the precision of gene editing with the targeting capabilities of peptides, illustrating the potential for highly personalized and effective treatments.

As research continues to unfold, the regulatory landscape for these experimental peptides will also evolve. Ensuring the safety and efficacy of new peptide therapies through rigorous clinical trials is paramount. The potential of peptides to target specific cellular mechanisms with minimal side effects holds great promise for the future of medicine, offering more effective, targeted, and personalized treatment options for a wide range of conditions. The ongoing exploration of experimental peptides not only expands our understanding of human biology but also brings us closer to curing diseases that remain a challenge today.

13.2: PEPTIDES IN PERSONALIZED MEDICINE

The integration of peptides into personalized medicine is a transformative approach that leverages the unique biochemical makeup of an individual to optimize health outcomes. This paradigm shift towards personalized peptide protocols is not merely about customizing treatments but about revolutionizing the way we approach health and disease management. The foundation of this approach is the understanding that each individual's body responds differently to peptides based on their genetic makeup, lifestyle, and existing health conditions. Therefore, the development of personalized peptide protocols involves a comprehensive evaluation of these factors to tailor treatments that are highly specific and effective for the individual.

Genetic Profiling plays a pivotal role in this process. By analyzing an individual's genome, healthcare providers can identify genetic variants that affect peptide metabolism, receptor sensitivity, and risk of adverse reactions. This information is crucial in selecting the most appropriate peptides and determining the optimal dosages to achieve the desired therapeutic effects while minimizing potential side effects. For instance, a person with a genetic predisposition for enhanced GHRH receptor sensitivity may experience better outcomes with lower doses of growth hormone-releasing peptides, thereby reducing the risk of side effects associated with higher dosages.

Lifestyle and Health Status Assessment is another critical aspect of developing personalized peptide protocols. Factors such as diet, exercise, stress levels, and sleep patterns can significantly influence the efficacy of peptide therapies. A thorough assessment of these factors allows for the adjustment of peptide protocols to complement the individual's lifestyle, thereby enhancing the therapeutic outcomes. For example, an individual with a high-stress lifestyle may benefit from peptides that have a calming effect on the nervous system, in addition to those targeting their primary health goals.

Continuous Monitoring and Adjustment of peptide protocols is essential in personalized medicine. The body's response to peptides can change over time due to various factors, including age, lifestyle changes, and health status fluctuations. Regular monitoring through blood tests, biomarker analysis, and feedback on subjective health measures enables healthcare providers to adjust peptide protocols as needed to ensure they remain effective and safe over the long term. This dynamic approach ensures that the personalized peptide protocol continues to align with the individual's evolving health needs and goals.

Collaboration Between Patients and Healthcare Providers is fundamental in the successful implementation of personalized peptide protocols. This collaborative approach involves open communication and shared decision-making, where patients are actively engaged in their treatment plans. Educating patients about the rationale behind their personalized peptide protocols, potential benefits, and risks, and the importance of adherence and lifestyle modifications empowers them to take an active role in their health optimization journey.

The future of personalized medicine with peptides is incredibly promising, with ongoing research and technological advancements continuously expanding our understanding and capabilities. The development of more sophisticated genetic testing, biomarker analysis, and peptide synthesis methods will further enhance the precision and effectiveness of personalized peptide protocols. As we move forward, the integration of peptides into personalized medicine holds the potential to not only improve health outcomes but also to transform the landscape of healthcare into one that is more proactive, predictive, and personalized.

13.3: ETHICAL AND LEGAL CONSIDERATIONS

The ethical and legal considerations surrounding the future use and regulation of peptides are complex and multifaceted, requiring a careful balance between innovation and safety. As we venture into this new frontier, it is imperative to address the ethical implications of peptide therapies, particularly in terms of accessibility, fairness, and the potential for misuse. **Accessibility** is a primary concern, as the benefits of peptide therapies should be available to all individuals, not just those who can afford them. This raises questions about the role of healthcare systems and insurance companies in covering these treatments, ensuring that advancements in peptide therapy do not exacerbate existing health inequities. **Fairness** also comes into play when considering how peptide therapies might be used to enhance human abilities, such as physical strength or cognitive function. The prospect of "designer peptides" for human enhancement has sparked a debate over what constitutes a fair use of these substances, particularly in competitive settings like sports or academia. The potential for **misuse** of peptide therapies, whether for performance enhancement or as a means of

Chapter 13:
Future Trends in Peptide Therapy

circumventing aging, necessitates stringent regulatory frameworks to govern their distribution and use.

On the legal front, the regulation of peptides presents its own set of challenges. The current regulatory landscape is a patchwork of laws that vary significantly from one jurisdiction to another, leading to confusion and inconsistency in how peptides are classified, manufactured, and sold. There is a pressing need for **harmonized regulations** that can keep pace with the rapid advancements in peptide science, ensuring that these therapies are safe, effective, and ethically distributed. This includes establishing clear guidelines for clinical trials, manufacturing practices, and marketing claims, as well as mechanisms for reporting and addressing adverse reactions. Furthermore, the legal system must grapple with the **intellectual property** aspects of peptide therapies, as patents play a crucial role in incentivizing research and development. However, overly broad patents could stifle innovation and restrict access to life-saving treatments, highlighting the need for a balanced approach to intellectual property rights in the peptide sector.

Data privacy is another critical issue, especially with the rise of personalized peptide protocols that rely on genetic and biomarker data. Protecting patient information from unauthorized access or misuse is paramount, requiring robust data security measures and clear consent protocols. Patients must be fully informed about how their data will be used, who will have access to it, and the measures in place to safeguard their privacy.

As we navigate these ethical and legal considerations, it is crucial to foster an ongoing dialogue among scientists, healthcare providers, regulators, and the public. This collaborative approach will help ensure that peptide therapies are developed and used in a manner that respects individual rights and societal values, paving the way for a future where these powerful tools can be harnessed for the greatest possible good.

Conclusion

SUMMARY OF KEY PROTOCOLS

The essence of peptide therapy lies in its ability to target and modulate biological processes with precision, offering a spectrum of benefits from enhancing muscle recovery to improving cognitive function. Key protocols include the use of **GHRP-2 and GHRP-6** for stimulating growth hormone release, thereby supporting muscle synthesis and recovery. These peptides, when combined with a regimen of proper nutrition and exercise, can significantly enhance athletic performance and facilitate faster recovery times. For cognitive enhancement, **Semax and Selank** have been highlighted for their neuroprotective properties and ability to improve focus and memory. These peptides work by modulating neurotransmitter levels and enhancing brain function, offering a non-invasive approach to boosting cognitive clarity.

In the realm of anti-aging, **Epithalon** stands out for its telomerase activation capability, which can extend the length of telomeres, thereby promoting cellular longevity and reducing the signs of aging. Similarly, **GHK-Cu** has been recognized for its role in skin health, promoting collagen production, and repairing tissue damage, which are crucial for maintaining a youthful appearance. The protocols for immune system support emphasize the use of **Thymosin Alpha-1**, known for its ability to modulate the immune response and enhance the body's ability to fight infections.

Each of these peptides requires careful consideration regarding dosage, timing, and cycling to maximize their efficacy while minimizing potential risks. It is imperative to source peptides from reputable suppliers to ensure purity and potency. Monitoring biomarkers and adjusting protocols in response to individual responses can further optimize outcomes, making peptide therapy a highly personalized approach to health and wellness.

The integration of peptides with lifestyle adjustments, such as stress management, sleep optimization, and dietary modifications, can amplify the benefits of peptide therapy. This holistic approach ensures that individuals not only address specific health concerns but also enhance their overall well-being, demonstrating the transformative potential of peptide protocols when applied thoughtfully and systematically.

MAINTAINING LONG-TERM HEALTH WITH PEPTIDE THERAPY

Maintaining long-term health with peptide therapy requires a **strategic and disciplined approach** to ensure that the benefits are not only achieved but sustained over time. It is crucial to **monitor your body's responses** closely and adjust dosages or protocols as needed. Regular check-ins with a healthcare provider who is knowledgeable about peptide therapy can provide invaluable guidance and adjustments to your regimen. This proactive approach allows for the **optimization of peptide effects**, ensuring that your body continues to respond positively without developing tolerance or adverse reactions.

Consistency in administration is key to maintaining the benefits of peptide therapy. Whether it's for **immune support, muscle recovery, cognitive enhancement,** or **anti-aging,** adhering to recommended dosages and schedules without unnecessary interruptions maximizes the therapeutic effects. However, it's also important to recognize when to cycle off peptides to give your body a rest and prevent desensitization to their effects. **Cycling strategies** should be tailored to individual needs and the specific peptides being used, with careful consideration given to the body's feedback and any changes in baseline health metrics.

Lifestyle factors play a significant role in enhancing and sustaining the benefits of peptide therapy. A diet rich in **nutrients that support peptide efficacy**, such as proteins, healthy fats, and antioxidants, can amplify the positive effects of peptides on the body. Adequate hydration and regular physical activity not only improve overall health but can also enhance the body's responsiveness to peptide therapy. **Sleep quality** is another critical factor; ensuring sufficient restorative sleep supports the body's healing and regeneration processes, which peptides often aim to optimize.

Stress management techniques such as meditation, yoga, or even simple breathing exercises can further improve the outcomes of peptide therapy. Chronic stress can hinder the body's ability to fully benefit from peptides, so incorporating practices that reduce stress levels is essential for long-term success.

Educating yourself about the nuances of peptide therapy is crucial. Stay informed about the latest research, emerging peptides, and their potential applications. This knowledge not only empowers you to make informed decisions about your health but also enables you to engage in meaningful discussions with your healthcare provider about adjusting your peptide therapy protocols as new information becomes available.

Finally, **tracking and documenting** your experience with peptide therapy can provide insights into its long-term effects on your health. Monitoring changes in your symptoms, performance, and overall well-being can help you and your healthcare provider make informed decisions about continuing, adjusting, or discontinuing certain peptides. This personalized approach ensures that peptide therapy remains a beneficial component of your health optimization strategy, tailored to your evolving needs and goals.

By following these strategies, individuals can maximize the long-term benefits of peptide therapy, ensuring that it remains an effective tool for health optimization and disease prevention. The key is to approach peptide therapy with patience, diligence, and a willingness to adapt protocols based on personal health outcomes and advancements in peptide research.

FINAL THOUGHTS AND NEXT STEPS

As you continue to explore the world of peptide therapy, it's essential to approach each step with a mindset geared towards **experimentation and adaptation**. The field of peptides is vast and ever-evolving, with new research and peptides emerging regularly. Staying informed about these developments can significantly impact the effectiveness of your peptide protocols. Engage with communities, forums, and publications dedicated to peptide research and biohacking. Sharing experiences and insights with others can provide new perspectives and strategies that you may not have considered.

Consulting with healthcare professionals who specialize in peptide therapy can provide a tailored approach to your regimen. These experts can offer insights into the latest peptide therapies and how they can be integrated into your health plan for optimal results. Remember, the goal is to use peptides in a way that supports your body's natural processes, enhancing your health without causing undue stress on your system.

Documenting your journey with peptides is invaluable. Keeping a detailed log of the peptides used, dosages, administration times, and any side effects or benefits noticed over time can help you and your healthcare provider make informed adjustments. This documentation can also serve as a valuable resource for the peptide community, contributing to the collective knowledge and understanding of peptide use.

As you adapt your peptide protocols, consider the broader aspects of your health and lifestyle. **Nutrition, exercise, sleep, and stress management** are foundational elements that can influence the effectiveness of peptides. Optimizing these areas of your life can enhance the benefits of peptide therapy, leading to improved outcomes.

Finally, remember that patience and persistence are key. The effects of peptides can sometimes be subtle or take time to become apparent. Regular monitoring and adjustments based on your body's responses are crucial for achieving the desired outcomes. With a thoughtful and informed approach, peptide therapy can be a powerful tool in your health optimization arsenal, offering benefits that extend far beyond the immediate goals of vitality, recovery, sleep, and focus.

Made in the USA
Las Vegas, NV
08 February 2025